Elementary
Algebra

Elementary Algebra

With Common Errors
and True-False Drill

Ronald J. Fischer

Archway Publishing books may be ordered through booksellers or by contacting:

Archway Publishing
1663 Liberty Drive
Bloomington, IN 47403
www.archwaypublishing.com
1 (888) 242-5904

ISBN: 978-1-4808-7712-2 (sc)
ISBN: 978-1-4808-7710-8 (hc)
ISBN: 978-1-4808-7711-5 (e)

Library of Congress Control Number: 2019906422

Print information available on the last page.

Archway Publishing rev. date: 7/16/2019

Preface

This book is a short review of elementary algebra. The student should be able to read it straight through in a short time.

Students of mathematics are often baffled by their own algebraic mistakes—and with good reason. The common manipulation errors of elementary algebra so closely resemble the actual laws that they can be called *nonlaws*. Students from algebra one through college algebra and calculus continue to make these deceptive errors.

The purpose of this little book is to aid the student in eliminating or reducing significantly the making of these common errors. It is intended for the student who has had some algebra, maybe even a lot of algebra, in the past but feels haunted by algebraic errors. Recent algebra texts do emphasize somewhat the common errors, but this emphasis is quite often light. In this book, mastering the common errors is the main goal. It is meant not as a textbook but as a supplement to any course that uses algebra.

Part 1 is a statement of the laws and common errors of elementary algebra. At the end of this part is a short summary of these laws and errors.

Part 2 is the heart of the book. It contains four drill tests consisting of a total of two hundred true-false and twenty multiple-choice questions for reinforcement. Classroom experience with this type of drill seems to perk up student interest significantly. Eyes seem to widen and the discussion livens as the differences between the law and its closely allied error become clear. The questions have been collected over a ten-year period and represent a comprehensive list. The student is urged to go over these questions two, three, or more times until his or her score is nearly perfect. (Grade "right minus wrong" to minimize the effects of guessing.) These questions are presented twice for the convenience of the student: first without solutions and second with complete solutions.

Contents

Part 1
The Laws and Common Errors

Introduction

These are the laws of arithmetic and the algebra of the real numbers. Numbers are represented by the usual decimals, fractions, and literals, which are the letters of the alphabet. Subscripts are often used to represent the first x, the second x (x_1, x_2) etc.

The real numbers can be represented on a horizontal line extending to infinity from the negative numbers to the left and the positive numbers to the right, with zero in the center. Any point on the line corresponds to a real number.

We use the usual symbols for addition (+) and subtraction (-). Note that the operation of multiplication is not usually written. So ab means a times b, and $2x$ means two times x. The exception is when we multiply two numbers, we must put something between them since, for example, 23 is interpreted as "twenty-three" and not "two times three." To indicate "two times three," we write it in one of the following ways: 2(3), (2)3, (2)(3), or 2•3. In the latter, the dot indicates multiplication. In algebra, we never use the x to represent multiplication since we often use x as a variable name.

The symbol for division may take several forms. We may write a/b, $\frac{a}{b}$, or $a \div b$. The latter is rarely used in algebra and will not be used in this book.

The following are the rules, also called the laws, that these numbers obey. We use the letters $a,\ b, c, \cdots, x, y, z$ to represent real numbers. Complex numbers will not be discussed.

1. The Commutative Laws of Addition and Multiplication

$$a + b = b + a \qquad\qquad ab = ba \qquad\qquad (1.1)$$

Note that these laws apply to addition and multiplication only, not any other operation, such as subtraction and division.

$$a - b \neq b - a \qquad\qquad a/b \neq b/a \qquad\qquad (1.2)$$

In fact, $a - b$ is the negative of $b - a$, and a/b is the reciprocal of b/a.

$$a - b = -(b - a) \qquad\qquad a/b = \frac{1}{b/a} \qquad\qquad (1.3)$$

2. The Associative Laws of Addition and Multiplication

The associative laws tell us rules about how we can add and multiply three numbers.

$$a + (b+c) = (a+b) + c = a + b + c \tag{2.1}$$

$$a(bc) = (ab)c = abc \tag{2.2}$$

Since the order of operation does not matter, we may omit the parentheses as in the third term of the equation. Note that these laws are only true for addition and multiplication and not subtraction or division. These rules can be generalized to any number of terms. In other words, addition and multiplication can be done in any order.

$$(2a)(b)cd(efg) = 2abcdefg$$

3. The Distributive Law

This law has to do with how multiplication and addition (or subtraction) relate to each other.

$$a(b+c) = ab + ac \text{ or } a(b-c) = ab - ac \tag{3.1}$$

Note that by the commutative law of multiplication $a(b+c) = (b+c)a$, so we can write (3.1) as follows:

$$(b+c)a = ba + ca \text{ or } (b-c)a = ba - ca \tag{3.2}$$

We say the multiplication distributes over addition or subtraction. It is not true that multiplication distributes over multiplication or division.

$$a(bc) \neq ab \cdot ac \text{ or } a(b/c) \neq (ab/ac) \tag{3.3}$$

To factor an expression means to write it as a product. We can use the distributive law in reverse to factor.

$$ab + ac = a(b+c) \tag{3.4}$$

The left-hand side is a *sum*, but the right-hand side is a *product*. The distributive law can be generalized to any number of terms.

$$a(b+c-d) = ab + ac - ad \tag{3.5}$$

A further generalization is to multiply two terms by two terms. Here we use the distributive law twice.

$$(a+b)(c+d) = (a+b)c + (a+b)d = ac + bc + ad + bd \tag{3.6}$$

Here the distributive law has been used once with $a+b$ and then again with c and d.

A common error is to distribute the first term only once.

$$a(b+c) \neq ab+c \tag{3.7}$$

This happens more often if the first quantity is a product of two or more terms.

$$ab(c+d) = abc+abd \text{ not } abc+bd \tag{3.8}$$

Another error occurs with fractions.

$$a \cdot \frac{1}{a}(c+d) \neq a \cdot \frac{1}{a}c+ad \tag{3.9}$$

But notice this:

$$a \cdot \frac{1}{a}(c+d) = 1(c+d) = c+d \tag{3.10}$$

See section (7.2).

4. The Additive Identity Law

The number 0 is called the *additive identity element*.

$$a + 0 = a \qquad (4.1)$$

Zero added to any number yields that same number. Zero is called the additive identity element. Of course, because of (1.1), the commutative law,

$$0 + a = a \qquad (4.2)$$

It is also true that

$$a - 0 = a \qquad (4.3)$$

But

$$0 - a \neq a \qquad (4.4)$$

In fact,

$$0 - a = -a \qquad (4.5)$$

which is the additive inverse of a (see law 6).

A rather obvious fact, but one that should be stated anyway, is this:

$$a \cdot 0 = 0 \qquad (4.6)$$

Any number times zero equals zero. Also,

$$\frac{0}{a} = 0 \text{ if } a \neq 0 \qquad (4.7)$$

Zero divided by any number (except 0) equals zero.

5. The Multiplicative Identity Law

The number 1 is called the *multiplicative identity element*.

$$a \cdot 1 = a \qquad (5.1)$$

One multiplied by any number gives that same number. Again because of law 1,

$$1 \cdot a = a \qquad (5.2)$$

6. The Additive Inverse Law

For each number a, there is another number, $-a$, such that

$$a + (-a) = 0 \text{ and } (-a) + a = 0 \qquad (6.1)$$

$-a$ is called the *additive inverse* or simply the *negative* of a. Note that if a is positive, $-a$ will be negative, but if a is negative, then $-a$ will be positive. In other words, the negative of a number may or may not be negative.

Also, zero is the only number that is equal to its own negative.

$$-0 = 0 \qquad (6.2)$$

The additive inverse of the additive inverse of a number is the original number. In symbols, it is this:

$$-(-a) = a.$$

The reason for this is as follows:

If $a + b = 0$, then $b = -a$.

Stated another (rather tricky) way is as follows:

If $-a + a = 0$, then $a = -(-a)$.

7. The Multiplicative Inverse Law

For each number a, *except zero*, there is another number a^{-1} such that

$$aa^{-1} = 1 \text{ and } a^{-1}a = 1 \tag{7.1}$$

a^{-1} is called the *multiplicative inverse* of a.

a^{-1} is also called the *reciprocal* of a and is written $1/a$. The above can then be written as such:

$$a \cdot \frac{1}{a} = 1 \tag{7.2}$$

It is important to realize that zero has no multiplicative inverse. Suppose it had one. Let us call it 0^{-1}. Then

$$0 \cdot 0^{-1} = 1.$$

But zero times anything must be zero.

$$0 \cdot 0^{-1} = 0$$

These last two equations would imply that one equals zero, which obviously cannot be true. We can only conclude that zero has no inverse. That is, 0^{-1} does not exist. This is related to a popular statement: "You can't divide by zero." (See appendix A.)

Stated another way,

$$\frac{a}{0} \text{ is not a number} \tag{7.3}$$

It is also true that the multiplicative inverse of the multiplicative inverse (the reciprocal of the reciprocal) of any number (except 0) is the number itself.

$$(a^{-1})^{-1} = a \ \text{ or } \ \frac{1}{1/a} = a \qquad\qquad \text{if } a \neq 0 \qquad\qquad (7.4)$$

Note that $a/0$ is not 0,

$$\frac{a}{0} \neq 0 \qquad\qquad (7.5)$$

a common error associated with (7.3).

8. The Definition of Subtraction

Subtraction is defined in terms of addition and the additive inverse.

$$a - b = a + (-b) \qquad (8.1)$$

That is, "a minus b," which means "Add to a the additive inverse (negative of) b."

From this and law 6, we can deduce the rather obvious fact that $a - a = 0$ as follows:

$$a - a = a + (-a) = 0 \qquad (8.2)$$

Also from (1.1),

$$-b + a = a + (-b) = a - b \qquad (8.3)$$

9. The Definition of Division

Division is defined in terms of multiplication and the multiplicative inverse (reciprocal). Three symbols for division are in common use: ÷, /, and the fraction line. In the interest of completeness, we give all three.

$$a \div b = a/b = \frac{a}{b} = ab^{-1} \tag{9.1}$$

That is "a divided by b" means
"multiply a by the multiplicative inverse of b."

From this and (7), we can easily deduce that

a/a = 1.

$$\frac{a}{a} = a \cdot \frac{1}{a} = aa^{-1} = 1 \tag{9.2}$$

Any number divided by itself equals one.

It is worth noting that the common symbol for division in computer languages is the slash (/). (See "Preface.") Hence, (9.2) can be written $a/a = 1$.

It also follows from (9.1) that

$$a/1 = \frac{a}{1} = a \cdot \frac{1}{1} = a \cdot 1 = a \tag{9.3}$$

Note that

$$\frac{1}{a} \neq a \tag{9.4}$$

and

$$\frac{1}{a} \neq -a \tag{9.5}$$

two common errors.

10. Sign Laws for Multiplication

$$(-a)b = -(ab) = -ab \tag{10.1}$$

$$a(-b) = -(ab) = -ab \tag{10.2}$$

$$(-a)(-b) = ab \tag{10.3}$$

These rules are often stated verbally as "minus times plus equals minus" and "minus times minus equals plus." In the third column of the first two lines above, $-ab$ is understood to be the same as $-(ab)$, that is the negative of ab.

In general, if we have an odd number of minuses multiplied together, the result is minus. If we have an even number, the result is plus. Thus,

$$(-a)(-b)c(-d) = -abcd \tag{10.4}$$

and

$$(-a)b(-c)cd = abcd \tag{10.5}$$

The following special case is worth noting:

$$(-1)a = -(1 \cdot a) = -a \tag{10.6}$$

Minus one times a number is its additive inverse (negative). Also from (10.6) and the distributive law,

$$-(a-b) = (-1)(a-b) = -a+b = b-a \tag{10.7}$$

and

$$a-(b-c) = a-b+c \tag{10.8}$$

See (2.3) for a common error associated with (10.8).

Generalizations are

$$-(a+b-c) = -a-b+c \qquad\qquad (10.9)$$

and

$$-(-a+b+c) = a-b-c \qquad\qquad (10.10)$$

This process is often called "removing parentheses." If there is a negative sign in front of a parenthesis, change the sign of everything in the parentheses.

An important special case of (10.8) is

$$a-(-b) = a+b \qquad\qquad (10.11)$$

"Subtracting a negative is the same as adding a positive."

Also from (10.7) we get an important rule for reducing fractions:

$$\frac{a-b}{b-a} = \frac{-(b-a)}{b-a} = -\frac{b-a}{b-a} = -1 \qquad\qquad (10.12)$$

Students often ask, "Why is minus times minus equal to plus?" This is often proven in advanced algebra books. For a convincing argument, see appendix B.

11. Sign Laws for Division

$$\frac{-a}{b} = -\frac{a}{b} \tag{11.1}$$

$$\frac{a}{-b} = -\frac{a}{b} \tag{11.2}$$

$$\frac{-a}{-b} = \frac{a}{b} \tag{11.3}$$

Verbally these rules say, "minus divided by plus equals minus," "plus divided by minus equals minus," and "minus divided by minus equals plus." In terms of the slash, the above can be written

$$(-a)/b = -(a/b) \tag{11.4}$$

$$a/(-b) = -(a/b) \tag{11.5}$$

$$(-a)/(-b) = a/b \tag{11.6}$$

In many books in applied math, the expression -a/b is seen. Does this mean (-a)/b or -(a/b)? According to (11.4), it does not matter, since both are the same.

The sign laws, (10.1)-(10.3) and (11.1)-(11.3) can be combined as follows:

$$\frac{(-a)(-b)c}{d(-e)} = -\frac{abc}{de} \tag{11.7}$$

$$\frac{(-a)(-b)c}{d(-e)(-f)} = \frac{abc}{def} \tag{11.8}$$

That is, an even number of minuses is plus, and an odd number of minuses gives minus when multiplying or dividing.

From (11.2) with $a = 1$ and $b = 1$,

$$\frac{-1}{1} = -\frac{1}{1} = -1 \tag{11.9}$$

17

or, since $\dfrac{1}{a} = a^{-1}$

$$\frac{1}{-1} = (-1)^{-1} = -1 \tag{11.10}$$

The multiplicative inverse of -1 is itself, -1.

A common error regarding (11.3) is

$$\frac{-a}{-b} \neq -\frac{a}{b} \tag{11.11}$$

12. Multiplication Law for Fractions

We now have the basic sign laws for real numbers. As in arithmetic, we now discuss the arithmetic of fractions—that is, how to add, subtract, multiply, and divide them. The rules are identical to those of arithmetic, except that the numbers may be negative. The rules for multiplication and division are simpler than addition and subtraction, so we discuss those first.

$$\frac{a}{b} \cdot \frac{c}{d} = \frac{ac}{bd} \tag{12.1}$$

To multiply two fractions, we simply multiply the numerators (tops) and multiply the denominators (bottoms). Using the slash, we can write the above law as

$$(a/b) \cdot (c/d) = (ac)/(bd) \tag{12.2}$$

The first form seems to be easier to understand.

This law is easily generalized to any number of fractions. Hence,

$$\frac{a}{b} \cdot \frac{c}{d} \cdot \frac{e}{f} = \frac{ace}{bdf} \tag{12.3}$$

Note that

$$\frac{a}{b} + \frac{c}{d} \neq \frac{a+c}{b+d} \tag{12.4}$$

a common error. The rule is for multiplication only, not addition. See law 17.

13. Division Law for Fractions

$$\frac{a}{b} \div \frac{c}{d} = \frac{a}{b} \cdot \frac{d}{c} = \frac{ad}{bc} \qquad (13.1)$$

To divide two fractions, we invert the divisor (right-hand fraction) and multiply. The last term is written by law 12. The above law can be written using the slash as

$$(a/b)/(c/d) = (a/b) \cdot (d/c) = (ad)/(bc) \qquad (13.2)$$

One more form is common,

$$\frac{a/b}{c/d} = \frac{a}{b} \cdot \frac{d}{c} = \frac{ad}{bc} \qquad (13.3)$$

Note that

$$\frac{a}{b} \div \frac{c}{d} \neq \frac{b}{a} \cdot \frac{c}{d} \qquad (13.4)$$

a common error. We invert the divisor (right side), not the dividend (left side).

14. Cancellation Law for Fractions

$$\frac{ac}{bd} = \frac{a}{b} \qquad (14.1)$$

Stated in words, "any number which is a common factor of both the top and bottom of a fraction can be canceled." We are actually dividing the top and bottom by c. This law is one of the most abused laws in elementary algebra. The numerator and denominator must be *products,* not *sums.* One of the most common errors is to cancel if it is part of a sum. For example,

$$\frac{a+c}{b+c} \neq \frac{a}{b} \qquad (14.2)$$

If you disagree with this, try letting $a = 1$, $b = 2$, $c = 3$. Then the left-hand side is

$$\frac{a+c}{b+c} = \frac{1+3}{2+3} = \frac{4}{5}$$

while the right-hand side is $\dfrac{a}{b} = \dfrac{1}{2}$.

Other examples of the misuse of the cancellation law are

$$\frac{a+c}{b+c} \neq \frac{a+1}{b+1} \text{ cancel the } c \qquad (14.3)$$

$$\frac{ab+c}{bd} \neq \frac{a+c}{d} \text{ cancel the } b \qquad (14.4)$$

and

$$\frac{a+b}{ac} \neq \frac{1+b}{c} \text{ cancel the } a. \qquad (14.5)$$

The list is nearly endless. The important thing to remember is that when canceling, both the top and bottom must be written as a *product.* If both top and bottom have the same factor, it may be canceled, even if it is a sum. For example,

$$\frac{a(b+c)}{d(b+c)} = \frac{a}{d}.$$

The left-hand side is expressed as a product. Numerator and denominator have the common factor, $b+c$, which can be canceled.

While we are not generally interested in proving things in this book, the proof of the cancellation law is interesting and quite simple.

Proof:

$$\frac{ac}{bc} = \frac{a}{b} \cdot \frac{c}{c} \text{ because of law 12}$$

$$= \frac{a}{b} \cdot 1 \text{ because of law 9}$$

$$= \frac{a}{b} \text{ because of law 5}$$

15. Building Law for Fractions

$$\frac{a}{b} = \frac{ac}{bc} \tag{15.1}$$

This law is the cancellation law in reverse. In reality, it is the same law. Stated in words, "given any fraction, we can multiply (or divide, but not add or subtract) both the top and the bottom of the fraction by the same number without changing the value of the fraction."

It may appear rather odd that by using this law, we are actually making a fraction more complicated, the reverse of simplifying. We need to do this in order to add or subtract fractions. (See section 17.)

Dividing the top and bottom of a fraction by the same quantity is also legal. In fact, dividing top and bottom by c is the same as multiplying top and bottom by $1/c$:

$$\frac{a}{b} = \frac{a \cdot (1/c)}{b \cdot (1/c)} = \frac{a/c}{b/c} \tag{15.2}$$

and hence it is the same law.

Two simple but important formulas worth noting are as follows:

$$\frac{ab}{b} = \frac{a}{1} = a \tag{15.3}$$

$$\frac{b}{ab} = \frac{1}{a} \tag{15.4}$$

Both of these are obtained by dividing top and bottom by b or canceling b. Watch out:

$$\frac{b}{ab} \neq a \tag{15.5}$$

Also note that since

$$\frac{a}{b} \cdot \frac{b}{a} = \frac{ab}{ba} = \frac{ab}{ab} = 1$$

a/b and b/a are reciprocals (multiplicative inverses) of each other. That is,

$$(a/b)^{-1} = \frac{1}{a/b} = \frac{b}{a} \qquad (15.6)$$

See law 7.

16. Adding and Subtracting Law for Fractions: Like Denominators

$$\frac{a}{c} + \frac{b}{c} = \frac{a+b}{c} \qquad (16.1)$$

$$\frac{a}{c} - \frac{b}{c} = \frac{a-b}{c} \qquad (16.2)$$

In words, "when adding or subtracting fractions that have the same denominator (bottom), add or subtract the numerators (tops), but leave the denominator the same."

We can generalize these laws as follows:

$$\frac{a}{c} + \frac{b}{c} - \frac{d}{c} = \frac{a+b-d}{c} \qquad (16.3)$$

or

$$\frac{a}{c} - \frac{b}{c} + \frac{d}{c} = \frac{a-b+d}{c} \qquad (16.4)$$

17. Adding and Subtracting Law for Fractions: Unlike Denominators

We will give this law in terms of a two-step procedure:

1. Use law 15 to build each fraction so that the denominators are the same.

2. Use law 16 to add or subtract.

Using this procedure to add

$$\frac{a}{b} + \frac{c}{d} \qquad (17.1)$$

we do the following:

1. $\dfrac{a}{b} = \dfrac{ad}{bd}$ and $\dfrac{c}{d} = \dfrac{cb}{db} = \dfrac{cb}{bd}$.

Now they have the same denominators. Therefore,

$$\frac{a}{b} + \frac{c}{d} = \frac{ad}{bd} + \frac{cb}{bd} \qquad (17.2)$$

Now the denominators are the same so that

2. $\dfrac{a}{b} + \dfrac{c}{d} = \dfrac{ad}{bd} + \dfrac{cb}{bd} = \dfrac{ad+cb}{bd} = \dfrac{ad+bc}{bd} \qquad (17.3)$

The result for subtraction is the same:

$$\frac{a}{b} - \frac{c}{d} = \frac{ad-bc}{bd} \qquad (17.4)$$

The term bd is called the common denominator of the two fractions, but it may not be the *least* common denominator.

Suppose the denominators have a factor in common. Let us call that factor f. Then

$$\frac{a}{bf}+\frac{c}{df}=\frac{a}{bf}\cdot\frac{d}{d}+\frac{c}{df}\cdot\frac{b}{b}$$

$$=\frac{ad+bc}{bdf} \tag{17.5}$$

That is, we use law 15 to build using only the noncommon parts of the denominators.

For subtraction, the rule is the same:

$$\frac{a}{bf}-\frac{c}{df}=\frac{ad-bc}{bdf} \tag{17.6}$$

We can generalize this to any number of fractions. For example,

$$\frac{a}{bh}+\frac{c}{dh}-\frac{f}{gh}=\frac{a}{bh}\cdot\frac{dg}{dg}+\frac{c}{dh}\cdot\frac{bg}{bg}-\frac{f}{gh}\cdot\frac{bd}{bd}$$

$$=\frac{adg+bcg-bdf}{bdgh} \tag{17.7}$$

Here, h is the factor common to all three denominators.

A special case of (17.3) is

$$a+\frac{c}{d}=\frac{a}{1}+\frac{c}{d}=\frac{ad}{d}+\frac{c}{d}=\frac{ad+c}{d} \tag{17.8}$$

Also,

$$a-\frac{c}{d}=\frac{a}{1}-\frac{c}{d}=\frac{ad-c}{d} \tag{17.9}$$

A common error associated with (17.8) and (17.9) is

$$a\frac{c}{d}\neq\frac{adc}{d} \tag{17.10}$$

but

$$a\frac{c}{d}=\frac{a}{1}\frac{c}{d}=\frac{ac}{d}$$

See (12.1).

18. Definition of Absolute Value

The absolute value or *magnitude* of a number, say b, is written $|b|$ and is defined as follows:

If b is positive or zero, then $|b| = b$ (18.1)

If b is negative or zero, then $|b| = -b$ (18.2)

Note that if b itself is negative, then $-b$ is positive, so that $|b|$ is always positive (or zero if $b = 0$). For example, if $b = 5$ then $|b| = 5$. If $b = -5$ then $|b| = -(-5) = 5$.

This rule seems unnecessarily complicated, and it is too complicated if we only use it for arithmetic calculations, but it is necessary for doing some algebraic problems.

As a consequence of this definition,

if $|x| = a$, then $x = \pm a$ (18.3)

In other words, if the absolute value of x is 5, then x can be 5 or minus 5. It depends on other facts to find out which it is.

19. Absolute Value Facts

$$|ab|=|a||b| \tag{19.1}$$

$$|a/b|=|a|/|b| \tag{19.2}$$

In words, "the absolute value of a product of two numbers equals the product of their absolute values," and "the absolute value of the quotient of two numbers equals the quotient of their absolute values."

The *triangle inequality* is

$$|a+b|\le|a|+|b| \tag{19.3}$$

In words, "the absolute value of the sum of two numbers is *less than or equal to* (which is written $\le$) the sum of their absolute values." (The name *triangle inequality* comes from the fact that every side of a triangle is less than the sum of the other two sides.)

And finally,

$$|a-b|\ge|a|-|b| \tag{19.4}$$

"The absolute value of the difference of two numbers is *greater than or equal to* (which is written $\ge$) the difference in their absolute values."

Note that in general,

$$|a+b|\ne|a|+|b| \tag{19.5}$$

and

$$|a-b|\ne|a|-|b| \tag{19.6}$$

which are common errors. $|a + b| = |a| + |b|$ if a and b are the same sign. $|a - b| = |a| - |b|$ if they are the same sign and $|a| \geq |b|$.

Laws (19.1) and (19.2) can be generalized to any number of terms, so for example,

$$\left|\frac{abc}{de}\right| = \frac{|a||b||c|}{|d||e|} \tag{19.7}$$

The triangle inequality (19.4) can also be generalized to

$$|a + b + c + d| \leq |a| + |b| + |c| + |d| \tag{19.8}$$

and (19.4) can be generalized to

$$|a - b - c - d| \geq |a| - |b| - |c| - |d| \tag{19.9}$$

20. Definition of Exponential Form

A number in the form x^n is said to be in exponential form. n is called the exponent, and x is called the base. If n is a positive whole number (positive integer), then

$$x^n = x \cdot x \cdot x \cdots x \qquad\qquad (20.1)$$

where there are n values of x on the right side. If the exponent is negative, say $-n$, then we define

$$x^{-n} = \frac{1}{x^n} \qquad\qquad (20.2)$$

For example,

$$2^3 = 2 \cdot 2 \cdot 2 = 8$$

$$2^{-3} = \frac{1}{2^3} = \frac{1}{8}$$

$$\left(\frac{2}{3}\right)^2 = (2/3)^2 = \frac{2}{3} \cdot \frac{2}{3} = \frac{2 \cdot 2}{3 \cdot 3} = \frac{4}{9}$$

$$(2.1)^2 = (2.1)(2.1) = 4.41$$

From law 10,

$$(-x)^2 = (-x)(-x) = x^2 \qquad\qquad (20.3)$$

$$(-x)^3 = (-x)(-x)(-x) = -(x)^3 = -x^3 \qquad\qquad (20.4)$$

In fact, if n is odd,

$$(-x)^n = -(x)^n = -x^n \qquad\qquad (20.5)$$

and if n is even,

$$(-x)^n = x^n \tag{20.6}$$

A common error in dealing with expressions of the form

$(-x)^n$ is

$$-3^2 = -9 \tag{20.7}$$

but

$$-3^2 \neq 9 \tag{20.8}$$

Powers are always performed before negations (see law 31). In general,

$$-x^n \neq (-x)^n \text{ but } -x^n = -(x^n) \tag{20.9}$$

21. Definition of Roots

Finding the nth root of a number, say b, is equivalent to finding a number whose nth power is b; that is, solving the equation,

$$x^n = b \tag{21.1}$$

for x. It is written $\sqrt[n]{b}$. If $n = 2$, the root n is not written so that the square root of 5 is written $\sqrt{5}$.

If n is even and $b \geq 0$ then,

$$x = \pm\sqrt[n]{b} \tag{21.2}$$

That is, there are two (real) nth roots of b. For example, if $x^2 = 9$, then $x = \pm\sqrt{9} = \pm 3$.

If n is odd, there is only one (real) nth root of b:

$$x = \sqrt[n]{b} \tag{21.3}$$

and it will have the same sign as b. For example, if

$x^3 = 27$, then $x = 3$. If $x^3 = -27$, then $x = -3$.

This is true because $3^3 = 27$ and $(-3)^3 = -27$.

If n is even, $\sqrt[n]{b}$ is called the principle or positive nth root of b. Hence, 3 is called the principle square root of 9. If b is negative and n is even, then no real roots exist; that is, $x = \sqrt[n]{b}$ is not a real number. It is an imaginary number, which is not covered in this book.

Note that

$$\sqrt{9} \neq \pm 3 \tag{21.4}$$

a common error. $\sqrt{9} = 3$ only, not -3.

If b is not a perfect nth power, then $\sqrt[n]{b}$ cannot

be known exactly. For example, in solving $x^2 = 5$, we would simply write $x = \pm\sqrt{5}$. $\sqrt{5}$ is approximately 2.2361, but it will never be known exactly. It is called an irrational number, one that can not be written as a fraction.

The symbol $\sqrt[n]{}$ is called the radical sign, and in law 34, we will discuss it in more detail.

22. Fractional Powers of the Form *m/n.*

(Also called *rational powers* from the word *ratio*.)

If the exponent is of the form $1/n$ where n is a positive whole number, then

$$x^{1/n} = \sqrt[n]{x} \qquad\qquad (22.1)$$

For example,

$$9^{1/2} = \sqrt{9} = 3$$

$$(-8)^{1/3} = \sqrt[3]{-8} = -2$$

$$(.04)^{1/2} = \sqrt{.04} = .2$$

If the exponent is of the form m/n, there are two equivalent forms:

$$x^{m/n} = \sqrt[n]{x^m} \qquad\qquad (22.2)$$

or

$$x^{m/n} = (\sqrt[n]{x})^m \qquad\qquad (22.3)$$

That is, the numerator, m, of the exponent is the power, the denominator, n, is the root. These last two equations say that one may perform the power operation first and root second or vice versa. For example,

$$27^{2/3} = \sqrt[3]{27^2} = \sqrt[3]{729} = 9$$

or

$$27^{2/3} = (\sqrt[3]{27})^2 = (3)^2 = 9.$$

23. The Laws of Exponents: Same Base

$$x^m x^n = x^{m+n} \qquad\qquad (23.1)$$

$$\frac{x^m}{x^n} = x^{m-n} \qquad\qquad (23.2)$$

When multiplying two exponential numbers having the same base, (x), add the exponents. When dividing them, subtract the exponents. See the following examples:

$$x^5 x^2 = x^{5+2} = x^7$$

$$\frac{x^5}{x^2} = x^{5-2} = x^3$$

$$\frac{x^2}{x^5} = x^{2-5} = x^{-3} = \frac{1}{x^3}$$

Equation (23.2) can be used to give a natural meaning of x to the zero power. If $m = n$,

$$\frac{x^n}{x^n} = 1.$$

But by (23.2),

$$\frac{x^n}{x^n} = x^{n-n} = x^0.$$

Therefore, we define

$$x^0 = 1 \quad \text{if } x \neq 0 \qquad\qquad (23.3)$$

x cannot equal zero here, since we would have a zero in the denominator of (23.2).

Also, by definition,

$$x^1 = x \qquad\qquad (23.4)$$

The first power of x is x itself.

Equations (23.1) and (23.2) can be generalized to

$$\frac{Ax^n x^m}{Bx^p x^q} = \frac{Ax^{n+m-p-q}}{B} \tag{23.5}$$

where A and B can be any expressions.

Note that

$$x^m x^n \neq x^{mn} \tag{23.6}$$

and

$$\frac{x^m}{x^n} \neq x^{m/n} \tag{23.7}$$

two common errors.

Also note that

$$(-1)^0 \neq -1 \text{ but } (-1)^0 = 1 \tag{23.8}$$

That is, minus 1 to the zero power = +1. Anything, positive or negative, (except 0) to the zero power equals 1.

24. The Laws of Exponents: Different Bases but Same Exponent

$$(xy)^n = x^n y^n \tag{24.1}$$

$$\left(\frac{x}{y}\right)^n = \frac{x^n}{y^n} \tag{24.2}$$

When raising the product of two numbers to a power, you may raise each to that power and multiply the results. When raising the quotient of two numbers to a power, you may raise each to that power and divide the results.

Note that,

$$(x+y)^n \neq x^n + y^n \tag{24.3}$$

and

$$(x-y)^n \neq x^n - y^n \tag{24.4}$$

These are very common errors. In other words, we cannot replace the multiplication and division operations above with addition or subtraction.

25. The Power to a Power Law

$$(x^m)^n = x^{mn} \tag{25.1}$$

When raising a number in exponential form to a power, you multiply the powers.

For example,

$$(x^2)^4 = x^{2 \cdot 4} = x^8$$

More generally,

$$(x^m y^n)^p = x^{mp} y^{np} \tag{25.2}$$

This law combines (24.1) with (25.1).

Note that

$$(x^m)^n \neq x^{m+n} \tag{25.3}$$

a common error.

26. Properties of Equality

$a = a$	reflexive property	(26.1)
If $a = b$, then $b = a$	symmetric property	(26.2)
If $a = b$ and $b = c$, then $a = c$	transitive property	(26.3)

27. Inequality Relations

The equal sign is the main *relation* in mathematics. Other important relations are

<   less than
>   greater than
≤   less than or equal to
≥   greater than or equal to

$$a < b \text{ means } b > a \tag{27.1}$$

$$a \leq b \text{ means } b \geq a \tag{27.2}$$

For the relations < and >, only the transitive property is true (see 26); the reflexive and symmetric are not:

$$\text{If } a < b \text{ and } b < c, \text{ then } a < c \text{ transitive property} \tag{27.3}$$

$$\text{If } a > b \text{ and } b > c, \text{ then } a > c \text{ transitive property} \tag{27.4}$$

For the relations ≤ and ≥, the reflexive and transitive properties are true, but the symmetric is not:

$$a \leq a \tag{27.5}$$

$$a \geq a \tag{27.6}$$

$$\text{If } a \leq b \text{ and } b \leq c, \text{ then } a \leq c \text{ transitive property} \tag{27.7}$$

$$\text{If } a \geq b \text{ and } b \geq c, \text{ then } a \geq c \text{ transitive property} \tag{27.8}$$

28. Equations and Rules for Solving Them

Suppose P and Q are both numbers or any algebraic expressions. Then $P = Q$ is called an equation.

If it is true that $P = Q$ and if c is any number, then it is true that

$$P + c = Q + c \tag{28.1}$$
$$P - c = Q - c \tag{28.2}$$
$$Pc = Qc \tag{28.3}$$
$$P/c = Q/c \quad \text{if } c \neq 0 \tag{28.4}$$

In other words, if you add, subtract, multiply, or divide both sides of an equation by any number, the equation remains true. The number may be zero in the first three cases (although the result would be pointless), but it cannot be zero in the fourth case, since division by zero is not allowed (see law 7 and appendix A).

One special case is worth noting.

$$\text{If } \frac{a}{b} = \frac{c}{d}, \text{ then } ad = bc \tag{28.5}$$

This is called *cross multiplication*, indicated schematically as

$$\frac{a}{b} \times \frac{c}{d}$$

This rule can be proven by simply multiplying both sides by bd and then canceling b on the left and d on the right, as follows:

$$\frac{a}{b}bd = \frac{c}{d}bd$$

$$ad = bc$$

See (15.3).

29. Inequalities and the Rules for Solving Them

If P and Q are both algebraic expressions, then $P < Q$, $P > Q$, $P \leq Q$, and $P \geq Q$ are called *inequalities*. The rules for solving them are almost the same as for solving equations (law 27). There is an exception, however.

If it is true that $P < Q$ and c is any number, then it is true that

$$P + c < Q + c \tag{29.1}$$

$$P - c < Q - c \tag{29.2}$$

That is, we may add or subtract the same number on both sides, just as with equations.

The rules for multiplication and division and slightly different and depend on whether c is positive or negative.

If $P < Q$ and $c > 0$ (c is positive), then

$$Pc < Qc \tag{29.3}$$

$$P/c < Q/c \tag{29.4}$$

However,

if $P < Q$ and $c < 0$ (that is, if c is negative), then

$$Pc > Qc \tag{29.5}$$

$$P/c > Q/c \tag{29.6}$$

That is, the inequality $<$ must be reversed to $>$. The reason for this becomes clear if we consider a simple example. We know that $-1 < 3$.

If we multiply both sides of this by -2, what is the inequality symbol that must be used?

2 ? -6

Obviously, it is 2>-6, indicating that the inequality symbol must be reversed.

In this section, in each of the above laws, we can replace < with ≤ and > with ≥. We can also replace each inequality symbol with its opposite. For example, if $P \geq Q$ and $c < 0$, then $Pc \leq Qc$.

30. The Substitution Principle

In any expression, any given quantity may be substituted for its equal. For example, if

$$(x-1)^2 + 3 = 2(x-1),$$

we can let $y = x - 1$ and write

$$y^2 + 3 = 2y.$$

This rule is used many times in both pure and applied mathematics.

The substitution principle can give new laws from old. For example, by letting $a = \dfrac{1}{d}$ in (3.2), $a(b \pm c) = ab \pm ac$, and using (9.1),

$$a(b \pm c) = ab \pm ac \text{ implies } \frac{1}{d}(b \pm c) = \frac{1}{d}b \pm \frac{1}{d}c = \frac{b}{d} \pm \frac{c}{d}.$$

Since this is a general law, the actual letters we use do not matter, and this is the same law as (3.4).

Also, if $b = 1$ and $c = 1$ in (3.2),

$$a(\frac{1}{e} \pm \frac{1}{f}) = a\frac{1}{e} \pm a\frac{1}{f},$$

which is the same law as (3.5).

31. Evaluation

In mathematics, one is often faced with the task of evaluating or calculating the value of some rather long, complicated expression, given the numerical values of the quantities involved in the expression. We must therefore know in what order to do the calculation.

The expression may or may not have grouping symbols, that is parentheses (), brackets [], or braces {}.

Part 1: We will first assume no grouping symbols are present. We will also assume that the expression consists of a sequence of numbers, letters, and operations that make sense. (This is sometimes referred to as a *legal string* in computer theory.) We will assume that the symbol for division is / but not the horizontal line, which may act as a grouping symbol (see 32).

In the following table, operations with the lowest order number are to be done first, etc. If two operations have the same order number, they are to be done in the order in which they appear from left to right. (This is called the left-to-right rule.) This order is the same as the order that such computer languages as BASIC, FORTRAN, PASCAL, and the C language use to evaluate expressions.

OPERATION	ORDER NUMBER
Exponentiation (raising to a power)	1
Negation	2
Multiplication	3
Division	3
Addition	4
Subtraction	4

Examples:

Suppose $a = 2$, $b = 3$, $c = 6$, then

1) $ab + c = (2)(3) + 6$ multiply before adding
   $= 6 + 6 = 12$.

2) $a + bc = 2 + (3)(6)$ multiply before adding
   $= 2 + 18 = 20$.

3) $2b^2 = 2(3)^2$ powers before multiplication
   $= 2(9) = 18$.

4) $a - b + c = 2 - 3 + 6$
   $= -1 + 6 = 5$ left-to-right rule.

5) $-c^2 = -6^2 = -36$ powers before negation, see (20.7).

6) $c / a / b = 6/2/3 = 3/3 = 1$ left-to-right rule.

7) $c / ab = 6/(2)(3) = [6/2]3$ left-to-right rule
   $= (3)(3) = 9$.

Note: This rule is used incorrectly in some math books and most physics books. In those books, a / bc is interpreted as $a /(bc)$, which is algebraically incorrect. However,

8) $c /(ab) = 6 /(2 \cdot 3) = 6 / 6 = 1$ do inside parentheses first.

Part 2: If parentheses or other grouping symbols are present, apply the same rules to expressions inside the grouping symbol, starting with the innermost grouping symbol and working outward. For example, for the same values of a, b, and c, $a = 2$, $b = 3$, $c = 6$, we evaluate

$$a + b(c - a) = 2 + 3(6 - 2) = 2 + 3(4) = 2 + 12 = 14.$$

A common error is to compute $a + b$ first.

$$a + b(c - a) \neq (2 + 3)(6 - 2) = 5(4) = 20$$

To take a more complex example, evaluate

$$80a - \{c - 2[c - 3a^2(b + 2a)]\}$$

$$80(2) - \{6 - 2[6 - 3 \cdot 2^2(3 + 2 \cdot 2)]\}$$

$$80(2) - \{6 - 2[6 - 3 \cdot 2^2(3 + 4)]\}$$

$$160 - \{6 - 2[6 - 3 \cdot 2^2(7)]\}$$

$$160 - \{6 - 2[6 - 3 \cdot 4(7)]\}$$

$$160 - \{6 - 2[6 - 12(7)]\}$$

$$160 - \{6 - 2[6 - 84]\}$$

$$160 - \{6 - 2[-78]\}$$

$$160 - \{6 + 156\}$$

$$160 - 162$$

$$-2$$

32. Implicit Grouping Symbols

The horizontal line division symbol, the absolute value symbol, and the radical sign are all implicit grouping symbols. That is,

$$\frac{a+b}{c+c} = \frac{(a+b)}{(c+d)} = (a+b)/(c+d) \tag{32.1}$$

$$|a+b| = |(a+b)| \tag{32.2}$$

The square root symbol is also a grouping symbol.

$$\sqrt{a+b} = \sqrt{(a+b)}$$

As is the nth root,

$$\sqrt[n]{a+b} = \sqrt[n]{(a+b)} \tag{32.3}$$

A common error involving (32.1) is omitting the implicit parentheses when adding or subtracting fractions:

$$\frac{3x}{a} - \frac{y-x}{a} = \frac{3x-(y-x)}{a} = \frac{3x-y+x}{a} = \frac{4x-y}{a} \tag{32.4}$$

which is correct, but

$$\frac{3x}{a} - \frac{y-x}{a} \neq \frac{3x-y-x}{a} = \frac{3x-y-x}{a} = \frac{2x-y}{a}.$$

The error here is not realizing that the fraction line is an implicit grouping symbol.

33. More on Exponents

The laws of exponents, (20) through (25), are very general, yet there are traps one must avoid.

If the bases, x and y, are positive, the laws of exponents are always true, even if the exponents are any positive or negative whole number or fraction. Here are a few examples, from (23.1):

$$x^{1/2}x^{-1/6} = x^{1/2-1/6} = x^{1/3} \tag{33.1}$$

From (25.1):

$$(x^{1/2})^6 = x^{(1/2)6} = x^3 \tag{33.2}$$

and

$$(x^2)^{1/4} = x^{2(1/4)} = x^{1/2} = \sqrt{x} \tag{33.3}$$

If the exponents are any positive or negative whole numbers, the laws of exponents are always true, even if the bases are negative.

Problems arise, however, when the bases are negative and the exponents are fractional. Sometimes the laws are true, and sometimes they are not.

Consider, for example, (33.1). Both $x^{1/2}$ and $x^{-1/6}$ do not exist when x is negative (since we have an even root of a negative number), but $x^{1/3}$ does exist.

Similarly for (33.2), when x is negative, $x^{1/2}$ does not exist, but x^3 does. In (33.3) the left side exists but the right sides do not. These problems are cleared up in the study of imaginary numbers. Here it is enough to be aware of them.

One more important case:

$$\sqrt{x^2} \neq x \text{ if } x \text{ is negative} \tag{33.4}$$

For example,

$$\sqrt{(-2)^2} \neq -2 \text{ but it equals } +2.$$

(Recall that $\sqrt{x}$ means the principal or positive square root of x, so it is always positive. See law 21.) It is correct to say, however, that

$$\sqrt{x^2} = |x| \qquad\qquad (33.5)$$

In fact, if n is even,

$$\sqrt[n]{x^n} = |x| \qquad\qquad (33.6)$$

and if n is odd,

$$\sqrt[n]{x^n} = x \qquad\qquad (33.7)$$

For example,

$$\sqrt[3]{(-5)^3} = \sqrt[3]{-125} = -5.$$

Similarly,

$$\sqrt{x^2 y^2} = |xy| \qquad\qquad (33.8),$$

but

$$\sqrt{x^2 y^2} = xy$$

only when xy is positive. That is when x and y are both positive or both negative. However,

$$\sqrt[3]{x^3 y^3} = xy \qquad\qquad (33.9)$$

is always true.

A very common error related to (33.8) and (33.9) is

$$\sqrt{x^2 + y^2} \neq x + y \qquad (33.10)$$

More generally,

$$\sqrt[n]{x^n + y^n} \neq x + y \qquad (33.11)$$

A careful examination of the laws of exponents reveals that they only apply to exponential numbers that are multiplied or divided, not added or subtracted.

The error of (33.10) can be easily seen by numerical example. If $x = 3$ and $y = 4$, the left-hand side of (33.10) is,

$$\sqrt{3^2 + 4^2} = \sqrt{9 + 16} = \sqrt{25} = 5$$

while the right-hand side is 3 + 4 = 7.

34. Roots as Exponents

The laws of roots (also called laws of radicals) are often stated separately from the laws of exponents, even though by law (22.1), a root is a special exponent. Therefore, for completeness, we will rewrite the laws of exponents (24.1), (24.2), and (25.1) in terms of roots, using $1/n$ and $1/m$ in place of n and m:

$$(xy)^{1/n} = x^{1/n}y^{1/n} \text{ is the same as } \sqrt[n]{xy} = \sqrt[n]{x}\sqrt[n]{y} \tag{34.1}$$

$$(x/y)^{1/n} = x^{1/n}/y^{1/n} \text{ is the same as } \sqrt[n]{\frac{x}{y}} = \frac{\sqrt[n]{x}}{\sqrt[n]{y}} \tag{34.2}$$

$$(x^{1/m})^{1/n} = x^{1/(mn)} \text{ is the same as } \sqrt[n]{\sqrt[m]{x}} = \sqrt[mn]{x} \tag{34.3}$$

The last rule can be proven as follows,

$$\sqrt[n]{\sqrt[m]{x}} = ((x)^{1/m})^{1/n} = x^{1/(mn)} = \sqrt[mn]{x}$$

where we are using (22.1) and (25.1).

35. Combining Like Terms

Law 3, the distributive law, can be used to simplify expressions that are added or subtracted. For example,

$$6x - 3x + 2x = (6 - 3 + 2)x = 5x$$

and

$$7a + 3b + 2a + 5bc - 7b + 3bc - 3a$$

$$= 7a + 2a - 3a + 3b - 7b + 5bc + 3bc$$

$$= (7 + 2 - 3)a + (3 - 7)b + (5 + 3)bc$$

$$= 6a - 4b + 8bc.$$

With practice, one can go from the first line to the last line by combining mentally like terms. Like terms are terms that have the identical literal (letter) part but not necessarily the same numerical part. As a final example,

$$= 3x^2 + 2x - 5x^2 + 3x = -2x^2 + 5x.$$

Note that the x^2 term and the x term are not like terms.

$$3x^2 + 2x \neq 5x^3 \text{ (or } 6x^3\text{)}$$

a common error that mixes up combining like terms with the laws of exponents.

36. Special Factors

The following formula come up often in algebra:

$$x^2 - y^2 = (x + y)(x - y) \qquad (36.1)$$

$$x^3 - y^3 = (x - y)(x^2 + xy + y^2) \qquad (36.2)$$

$$x^3 + y^3 = (x + y)(x^2 - xy + y^2) \qquad (36.3)$$

$$(x + y)^2 = x^2 + 2xy + y^2 \qquad (36.4)$$

$$(x - y)^2 = x^2 - 2xy + y^2 \qquad (36.5)$$

$$(x + y)^3 = x^3 + 3x^2y + 3xy^2 + y^3 \qquad (36.6)$$

$$(x - y)^3 = x^3 - 3x^2y + 3xy^2 - y^3 \qquad (36.7)$$

These formulas can all be proven by multiplying out the right (or left in the case of 36.4 to 36.7) hand side using the distributive law, law 3. For example, for (36.2),

$$(x - y)(x^2 + xy + y^2) = x(x^2 + xy + y^2) - y(x^2 + xy + y^2)$$

$$= (x^3 + x^2y + xy^2) - (x^2y + xy^2 + y^3)$$

$$= x^3 + x^2y + xy^2 - x^2y - xy^2 - y^3$$

$$= x^3 - y^3;$$

and for (36.7),

$$(x - y)^3 = (x - y)(x - y)^2 = (x - y)(x^2 - 2xy + y^2)$$

$$= x(x^2 - 2xy + y^2) - y(x^2 - 2xy + y^2)$$

$$= x^3 - 2x^2y + xy^2 - x^2y + 2xy^2 - y^3$$

$$= x^3 - 3x^2y + 3xy^2 - y^3$$

where in the fourth line of this example we have used law 35 to combine like terms.

Common errors to watch out for are as follows:

$$(x+y)^n \neq x^n + y^n \tag{36.8}$$

and

$$(x^n + y^n)^m \neq x^{nm} + y^{nm} \tag{36.9}$$

Some special cases of these are as follows:

$$(x+y)^2 \neq x^2 + y^2 \tag{36.10}$$

$$(x^2 + y^2)^3 \neq x^6 + y^6 \tag{36.11}$$

$$(x^{-1} + y^{-1})^{-1} \neq x + y \tag{36.12}$$

If all the pluses were times, these would be true. See (25.2).

Also,

$$(x-y)^2 = (y-x)^2 \tag{36.13}$$

but

$$(x-y)^3 = -(y-x)^3 \tag{36.14}$$

In general, if n is even,

$$(x-y)^n = (y-x)^n \tag{36.15}$$

but if n is odd,

$$(x-y)^n = -(y-x)^n \qquad (36.16)$$

These can be deduced using (10.7), (20.5), and (20.6).

The error (36.12) in fractional form is

$$\frac{1}{\dfrac{1}{x}+\dfrac{1}{y}} \neq x+y \qquad (36.17)$$

This error is made because of confusion with

$$\frac{1}{\dfrac{1}{x}\dfrac{1}{y}} = xy$$

which is true.

Finally, it is shown in advanced algebra courses that $x^n + y^n$ can never be factored if n is a power of 2. That is, if n is (2, 4, 8, 16, etc).

For example, $x^2 + y^2$ cannot be factored; nor can $x^8 + y^8$.

Note, however, that $x^6 + y^6$ can be factored, since it can be written as the sum of two cubes, $x^6 + y^6 = (x^2)^3 + (y^2)^3$ and we can use (36.3) to factor it.

$$x^6 + y^6 = (x^2)^3 + (y^2)^3 = (x^2 + y^2)(x^4 - x^2 y^2 + y^4)$$

37. Linear Equations in One Variable

This type of equation is called an equation of the first degree. All powers of x are 1. It can eventually always be put in the form

$$ax = b \qquad\qquad (37.1)$$

where a and b are known and for which x is to be solved. The rules for solving them are given in law 28. The general methods for solving them are given below.

1) If the equation contains fractions, use (28.3) to multiply both sides of the equation by the least common denominator (LCD). This will remove all fractions.

2) Remove all grouping symbols (parentheses, brackets, or braces) by using law 3, the distributive law.

3) Use law 35 to combine like terms at this step and whenever possible.

4) Use (28.1) and (28.2) to add or subtract terms to both sides of the equation in order to collect terms that contain the unknown variable (usually x) on one side of the equation and all other terms on the other side.

5) Divide both sides by the numerical coefficient of the unknown quantity. One side of the equation should contain the unknown variable and the other the numerical answer.

6) Check your answer by substituting it into the original equation (*not any other equation*) and evaluating each side using law 31. If the sides are equal, your answer is correct.

A common error is to use an equation other than the original one to check your answer. Always go back to the first one.

Example:

Solve the following for x:

$$\frac{x}{2}+\frac{x-3}{3}+1=\frac{2}{5}(x-5)+28$$

Solution:

Using the steps above,

1) Multiply both sides by 30, the least common denominator of 2, 3 and 5:

$$30\frac{x}{2}+30\frac{x-3}{3}+30\cdot1=30\frac{2}{5}(x-5)+30\cdot28$$

$$=15x+10(x-3)+30=12(x-5)+30\cdot28$$

2) Remove parentheses using law 3:

$$15x+10x-30+30=12x-60+840$$

3) Combine like terms:

$$25x=12x+780$$

4) Subtract 12x from both sides:

$$25x-12x=12x+780-12x$$

$$13x=780$$

5) Divide both sides by 13, giving the solution:

$$\frac{13x}{13}=\frac{780}{13}$$

$$x=60$$

6) Check your answer in *the original equation*. The left-hand side is

$$\frac{60}{2} + \frac{60-3}{3} + 1 = 30 + \frac{57}{3} + 1 = 30 + 19 + 1 = 50.$$

The right-hand side is

$$\frac{2}{5}(60-5) + 28 = \frac{2}{5}55 + 28 = 22 + 28 = 50.$$

Both sides are equal, so the answer checks.

38. The Literal Linear Equation

The literal equation is one that has other literals (or letters) in it besides the unknown quantity to be solved for. The rules for solving them are exactly the same as given in 37.

Example: Solve for x:

$$x - c = \frac{x+3}{b}$$

Solution:

1) Multiply both sides by b:

$$b(x-c) = b\frac{x+3}{b} = x+3$$

2) Remove grouping symbols:

$$bx - bc = x + 3$$

3) No like terms to combine.

4) Subtract x and add bc to both sides:

$$bx - x = bc + 3$$

5) With linear equations, often like terms cannot be combined, but by factoring out x, the effect is the same:

$$(b-1)x = bc + 3$$

Now divide both sides by $b-1$:

$$\frac{(b-1)x}{b-1} = \frac{bc+3}{b-1}$$

$$x = \frac{bc+3}{b-1}$$

6) Checking literal equations can sometimes be involved. The left-hand side is as follows:

$$\frac{bc+3}{b-1} - c = \frac{bc+3}{b-1} - \frac{c(b-1)}{b-1} = \frac{bc+3-bc+c}{b-1}$$

$$= \frac{3+c}{b-1} = \frac{c+3}{b-1}$$

The right-hand side is as follows:

$$\frac{\frac{bc+3}{b-1}+3}{b} = \frac{\frac{bc+3}{b-1}+\frac{3(b-1)}{b-1}}{b} = \frac{\frac{bc+3}{b-1}+\frac{3b-3)}{b-1}}{b}$$

$$= \frac{1}{b} \cdot \frac{bc+3b}{b-1} = \frac{bc+3b}{b(b-1)} = \frac{b(c+3)}{b(b-1)} = \frac{c+3}{b-1}$$

Both sides are equal, so the answer checks.

A noteworthy fact is that b cannot equal 1, since the denominator will be zero. If $b = 1$, we go back to the original equation and see what happens.

$$x - c = \frac{x+3}{1} = x+3$$

Here we can subtract x from both sides, and we have $-c = 3$. We no longer have an equation involving x, so there is no solution if $b = 1$. However, if $c = -3$, we have $x + 3 = x + 3$ and the solution is all real numbers.

39. The Quadratic Equation

The equation

$$ax^2 + bx + c = 0 \qquad (39.1)$$

where $a \neq 0$ (if a = 0, it is linear) is called a quadratic equation in x where a, b, and c are assumed to be given numbers. We will consider three cases:

Case 1, $b = 0$. The equation then becomes

$$ax^2 + c = 0 \qquad (39.2)$$

Then

$$ax^2 = -c, \; x^2 = \frac{-c}{a}$$
$$x = \pm\sqrt{\frac{-c}{a}} \qquad (39.3)$$

where we have used equation (21.2). Here for solutions to exist, a and c must be opposite in sign, so that $-c/a$ will be positive.

Case 2, $c = 0$. The equation then becomes

$$ax^2 + bx = 0 \qquad (39.4)$$

Using the distributive law, this can be written,

$$x(ax + b) = 0 \qquad (39.5)$$

Now if the product of two or more numbers or expressions is zero, at least one of them must be zero. This is rather obvious since if none of them is zero, the product cannot possibly be zero. We state this formally as follows:

If $PQR = 0$ then $P = 0$ or $Q=0$ or $R = 0$ $\hspace{2cm}$ (39.6)

We can apply this rule to (39.5) so that

$x = 0$ or $ax + b = 0$.

The second equation is linear, which can be solved as follows:

$ax + b = 0$

$ax = -b$

$x = \dfrac{-b}{a}$

Therefore, the two solutions of (39.5) are as follows:

$x = 0$ and $x = \dfrac{-b}{a}$ $\hspace{2cm}$ (39.7)

Before considering the general case, let us consider some examples.

Example 1. Solve $9x^2 - 2 = 0$. From case 1,

$9x^2 = 2$

$x^2 = \dfrac{2}{9}$

$x = \pm\sqrt{\dfrac{2}{9}} = \pm\dfrac{\sqrt{2}}{\sqrt{9}} = \pm\dfrac{\sqrt{2}}{3}.$

Example 2. Solve $3x^2 - 2x = 0$. This is a special case of (39.5), so

$x(3x - 2) = 0$

$x = 0$ or $3x - 2 = 0$

$x = 0$ or $x = \dfrac{2}{3}.$

Case 3, the general case. The equation (39.1) can be solved even if neither b nor c is zero. The solution is as follows:

$$x = \frac{-b \pm \sqrt{b^2 - 4ac}}{2a} \tag{39.8}$$

For a proof of this, multiply both sides by $2a$:

$$2ax + b = \pm\sqrt{b^2 - 4ac}$$

Now squaring both sides,

$$(2ax + b)^2 = b^2 - 4ac$$

$$4a^2x^2 + 4abx + b^2 = b^2 - 4ac.$$

The b^2 cancels, so we have

$$4a^2x^2 + 4abx + 4ac = 0.$$

And dividing both sides by $4a$, we get the original equation,

$$ax^2 + bx + c = 0 \text{ which is (39.1).}$$

These steps can be done in the reverse order to end up with equation (39.8).

As an example in solving the equation,

$$x^2 - 4x + 1 = 0,$$

$a = 1$, $b = -4$, and $c = 1$. Its solution is therefore

$$x = \frac{-(-4) \pm \sqrt{(-4)^2 - 4 \cdot 1 \cdot 1}}{2 \cdot 1} = \frac{4 \pm \sqrt{16 - 4}}{2} = \frac{4 \pm \sqrt{12}}{2}$$

$$= \frac{4 \pm \sqrt{4 \cdot 3}}{2} = \frac{4 \pm 2\sqrt{3}}{2} = \frac{2(2 \pm \sqrt{3})}{2} = 2 \pm \sqrt{3}.$$

The quantity $b^2 - 4ac$ is called the discriminant of the equation and must be positive or zero for real solutions to exist.

We can check our solution by substituting it into the original equation,

$$x^2 - 4x + 1 = (2 \pm \sqrt{3})^2 - 4(2 \pm \sqrt{3}) + 1$$

$$= 4 \pm 4\sqrt{3} + 3 - (8 \pm 4\sqrt{3}) + 1 = 4 + 3 - 8 + 1 = 7 - 8 + 1 = -1 + 1 = 0.$$

The result is zero, so it checks.

Try this problem: solve for x. $3x^2 + 5x - 2 = 0$.

Answer: $x = 1/3, \quad x = -2$.

40. Summary of Main Formulas

$$a+b=b+a \quad ab=ba \tag{1.1}$$

$$a+(b+c)=(a+b)+c=a+b+c \tag{2.1}$$

$$a(bc)=(ab)c=abc \tag{2.2}$$

$$a(b\pm c)=ab\pm ac \tag{3.1}$$

$$a(b+c-d)=ab+ac-ad \tag{3.7}$$

$$(a+b)(c-d)=ac+bc-ad-bd \tag{3.8}$$

$$a\pm 0=a \tag{4.1}, (4.3}$$

$$a\cdot 0=0 \tag{4.6}$$

$$\frac{0}{a}=0 \tag{4.7}$$

$$-(-a)=a \tag{4.8}$$

$$a\cdot 1=a \tag{5.1}$$

$$a+(-a)=0 \tag{6.1}, (8.2}$$

$$-0=0 \tag{6.2}$$

$$aa^{-1}=a\cdot\frac{1}{a}=\frac{a}{a}=1 \tag{7.1}, (7.2), (9.2}$$

$$\frac{a}{0} \text{ does not exist} \tag{7.3}$$

$$(a^{-1})^{-1}=\frac{1}{1/a}=a \tag{7.4}$$

$$a-b=a+(-b) \tag{8.1}$$

$$a - a = 0 \tag{8.2}$$

$$-b + a = a - b \tag{8.3}$$

$$a \div b = a / b = \frac{a}{b} = ab^{-1} = a \cdot \frac{1}{b} \tag{9.1}$$

$$\frac{a}{1} = a \tag{9.3}$$

$$(-a)b = a(-b) = -(ab) = -ab \tag{10.1}, (10.2)$$

$$(-a)(-b) = ab \tag{10.3}$$

$$(-1)a = -a \tag{10.6}$$

$$-(b - a) = a - b \tag{10.7}$$

$$a - (b - c) = a - b + c \tag{10.8}$$

$$-(a + b - c) = -a - b + c \tag{10.9}$$

$$a - (-b) = a + b \tag{10.11}$$

$$\frac{a - b}{b - a} = -1 \tag{10.12}$$

$$\frac{-a}{b} = \frac{a}{-b} = -\frac{a}{b} \tag{11.1}, (11.2)$$

$$\frac{-a}{-b} = \frac{a}{b} \tag{11.3}$$

$$\frac{a}{b} \cdot \frac{c}{d} = \frac{ac}{bd} \tag{12.1}$$

$$\frac{a}{b} \cdot \frac{c}{d} \cdot \frac{e}{f} = \frac{ace}{bdf} \tag{12.3}$$

$$\frac{a}{b} \div \frac{c}{d} = (a/b)/(c/d) = \frac{\frac{a}{b}}{\frac{c}{d}} = \frac{ad}{bc} \tag{13.1}, (13.3)$$

$$\frac{ac}{bc} = \frac{a}{b} \tag{14.1}, (15.1)$$

$$\frac{ab}{b} = \frac{a}{1} = a \tag{15.3}$$

$$\frac{b}{ab} = \frac{1}{a} \tag{15.4}$$

$$\left(\frac{a}{b}\right)^{-1} = \frac{1}{a/b} = \frac{b}{a} \tag{15.6}$$

$$\frac{a}{c} \pm \frac{b}{c} = \frac{a \pm b}{c} \tag{16.1, 16.2}$$

$$\frac{a}{c} + \frac{b}{c} - \frac{d}{c} = \frac{a + b - d}{c} \tag{16.3}$$

$$\frac{a}{b} \pm \frac{c}{d} = \frac{ad \pm bc}{bd} \tag{17.3, 17.4}$$

$$\frac{a}{bf} \pm \frac{c}{df} = \frac{ad \pm bc}{bdf} \tag{17.5, 17.6}$$

$$a \pm \frac{c}{d} = \frac{ad \pm c}{d} \tag{17.8, 17.9}$$

If $b \geq 0$, then $|b| = b$ \hfill (18.1)

If $b < 0$, then $|b| = -b$ \hfill (18.2)

$$|ab| = |a| \cdot |b| \tag{19.1}$$

$$\left|\frac{a}{b}\right| = \frac{|a|}{|b|} \tag{19.2}$$

$$|a + b| \leq |a| + |b| \tag{19.3}$$

$$|a - b| \geq |a| - |b| \tag{19.4}$$

$x^n = xxx \cdots x$ (n x's) if n is a positive integer \hfill (20.1)

$$x^{-n} = \frac{1}{x^n} \tag{20.2}$$

If n is odd, $(-x)^n = -x^n$ \hfill (20.5)

If n is even, $(-x)^n = x^n$ \hfill (20.6)

$x = \sqrt[n]{b}$ means $x^n = b$ (21.1), (21.3)

$x^{1/n} = \sqrt[n]{x}$ (22.1)

$x^{m/n} = \sqrt[n]{x^m}$ (22.2), (22.3)

$x^m x^n = x^{m+n}$ (23.1)

$\dfrac{x^m}{x^n} = x^{m-n}$ (23.2)

$x^0 = 1$ if $x \neq 0$ (23.3)

$(xy)^n = x^n y^n$ (24.1)

$\left(\dfrac{x}{y}\right)^n = \dfrac{x^n}{y^n}$ (24.2)

$(x^m)^n = x^{mn}$ (25.1)

Laws of equality:

    Symmetric: $a = a$ (26.1)

    Reflexive: If $a = b$, then $b = a$ (26.2)

    Transitive: If $a = b$ and $b = c$, then $a = c$ (26.3)

If $a < b$ and $b < c$, then $a < c$ (27.3)

If $P = Q$ then,

    1. $P \pm c = Q \pm c$ (28.1), (28.2)

    2. $Pc = Qc$ (28.3)

    3. $\dfrac{P}{c} = \dfrac{Q}{c}$ (if $c \neq 0$)   (28.4)

If $\dfrac{a}{b} = \dfrac{c}{d}$, then $ad = bc$ $\hfill$ (28.5)

If $P < Q$, then

1. $P \pm c < Q \pm c$ $\hfill$ (29.1), (29.2)

2. $Pc < Qc$ if $c > 0$ $\hfill$ (29.3)

3. $\dfrac{P}{c} < \dfrac{Q}{c}$ if $c > 0$ $\hfill$ (29.4)

4. $Pc > Qc$ if $c < 0$ $\hfill$ (29.5)

5. $\dfrac{P}{c} > \dfrac{Q}{c}$ if $c < 0$ $\hfill$ (29.6)

Substitution Principle:
Any quantity may be substituted for its equal. $\hfill$ (30)

$\sqrt[n]{x^n} = |x|$ if n is an even integer $\hfill$ (33.6)

$\sqrt[n]{x^n} = x$ if n is an odd integer $\hfill$ (33.7)

$\sqrt[n]{xy} = \sqrt[n]{xy} = \sqrt[n]{x}\sqrt[n]{y}$ $\hfill$ (34.1)

$\sqrt[n]{\dfrac{x}{y}} = \dfrac{\sqrt[n]{x}}{\sqrt[n]{y}}$ $\hfill$ (34.2)

$\sqrt[n]{\sqrt[m]{x}} = \sqrt[mn]{x}$ $\hfill$ (34.3)

$x^2 - y^2 = (x+y)(x-y)$ $\hfill$ (36.1)

$x^3 - y^3 = (x-y)(x^2 + xy + y^2)$ $\hfill$ (36.2)

$x^3 + y^3 = (x+y)(x^2 - xy + y^2)$ $\hfill$ (36.3)

$x^2 + 2xy + y^2 = (x+y)^2$ $\hfill$ (36.4)

$x^2 - 2xy + y^2 = (x-y)^2$ $\hfill$ (36.5)

$$(x+y)^3 = x^3 + 3x^2y + 3xy^2 + y^3 \qquad (36.6)$$

$$(x-y)^3 = x^3 - 3x^2y + 3xy^2 - y^3 \qquad (36.7)$$

The solution of the quadratic equation:

$$ax^2 + bx + c = 0 \qquad (39.1)$$

is the quadratic formula,

$$x = \frac{-b \pm \sqrt{b^2 - 4ac}}{2a} \qquad (39.8)$$

41. Summary of Common Errors

$$a - b \neq b - a \quad a/b \neq b/a \tag{1.2}$$

$$a - (b - c) \neq (a - b) - c \tag{2.3}$$

$$(a/b)/c \neq a/(b/c) \tag{2.4}$$

$$a(bc) \neq (ab)(ac) \tag{3.3}$$

$$a\frac{b}{c} \neq \frac{ab}{ac} \tag{3.3}$$

$$a(b \pm c) \neq ab \pm c \tag{3.9}$$

$$ab(c + d) \neq abc + ad \tag{3.10}$$

$$a \cdot \frac{1}{a}(c + d) \neq a \cdot \frac{1}{a}c + ad = c + ad \tag{3.11}$$

$$0 - a \neq a \tag{4.4}$$

$$\frac{a}{0} \neq 0 \tag{7.5}$$

$$\frac{1}{a} \neq a \tag{9.4}$$

$$\frac{1}{a} = a^{-1} \neq -a \tag{9.5}$$

$$\frac{-a}{-b} \neq -\frac{a}{b} \tag{11.11}$$

$$\frac{a + c}{b + c} \neq \frac{a}{b} \tag{14.2}$$

$$\frac{a + c}{b + c} \neq \frac{a + 1}{b + 1} \tag{14.3}$$

$$\frac{ab + c}{bd} \neq \frac{a + c}{d} \tag{14.4}$$

$$\frac{a + b}{ac} \neq \frac{1 + b}{c} \tag{14.5}$$

73

$$\frac{b}{ab} \neq a \tag{15.5}$$

$$a\frac{c}{d} \neq \frac{adc}{d} \tag{17.10}$$

$$|a+b| \neq |a|+|b| \tag{19.5}$$

$$|a-b| \neq |a|-|b| \tag{19.6}$$

$$-x^n \neq x^n \tag{20.9}$$

$$\sqrt{9} \neq \pm 3 \tag{21.4}$$

$$x^m x^n \neq x^{mn} \tag{23.6}$$

$$x^m / x^n = \frac{x^m}{x^n} \neq x^{m/n} \tag{23.7}$$

$$(x^m)^n \neq x^{m+n} \tag{25.3}$$

$$\frac{x}{a} - \frac{y-z}{a} \neq \frac{x-y-z}{a} \tag{32.4}$$

$$\sqrt{x^2} \neq x \text{ if } x < 0 \tag{33.4}$$

$$\sqrt{x^2+y^2} \neq x+y \tag{33.11}$$

$$\sqrt[n]{x^n+y} \neq x+\sqrt[n]{y} \tag{34.7}$$

$$(x+y)^n \neq x^n+y^n \tag{36.8}$$

$$(x^2+y^3)^2 \neq x^4+y^6 \tag{36.11}$$

$$(x^{-1}+y^{-1})^{-1} \neq x+y \tag{36.12}$$

$$(x-y)^3 \neq (y-x)^3 \tag{36.13}$$

$$\frac{1}{\frac{1}{x}+\frac{1}{y}} \neq x+y \tag{36.19}$$

Part 2
Drill Questions

Introduction

Part 2 is really the main point of this book. Most students who have studied some algebra are already familiar with part 1. The questions of part 2 will make clear the truth of algebra from the fiction or common errors of the subject. To obtain an accurate measure of your knowledge of algebra, grade yourself right minus wrong. Go over these questions again and again until your score is nearly perfect.

Drill 1A—General Principles, Signs, and Fractions

Answer each question true or false. (100 questions)

1. $(a+b)+c = a+(b+c)$

2. $a+b = b+a$

3. $ab = ba$

4. $a-b = b-a$

5. $a-(b-c) = (a-b)-c$

6. $(a/b)/c = a/(b/c)$

7. $a/b = b/a$

8. $a(b+c) = ab+ac$

9. $a(bc) = (ab)c$

10. $-b$ is never a positive number

11. $\dfrac{0}{2} = 0$

12. $3(ab) = (3a)(3b)$

13. $\dfrac{ac}{bc} = \dfrac{a}{b}$

14. $\dfrac{2}{0} = 0$

15. $2+3(6) = 30$

16. $\dfrac{a}{c} + \dfrac{b}{c} = \dfrac{a+b}{c}$

17. $\dfrac{a+c}{b+c}$ simplifies to $\dfrac{a}{b}$

18. $\dfrac{a+c}{b+c} = \dfrac{a+1}{b+1}$

19. $\dfrac{a+b}{ac} = \dfrac{1+b}{c}$

20. $\dfrac{3+x}{12}$ simplifies to $\dfrac{x}{4}$

21. $\dfrac{a}{c} - \dfrac{b}{c} = \dfrac{a-b}{c}$

22. $\dfrac{a}{b} \cdot \dfrac{c}{d} = \dfrac{ac}{bd}$

23. $\dfrac{a}{c} + \dfrac{b}{d} = \dfrac{a+b}{c+d}$

24. $\dfrac{a}{b} \div \dfrac{c}{d} = \dfrac{a}{b} \cdot \dfrac{d}{c}$

25. $\dfrac{a}{b} - \dfrac{c}{d} = \dfrac{a-d}{b-c}$

26. The fraction $\dfrac{a}{b}$ can be built to $\dfrac{ac}{bc}$

27. $\dfrac{a}{b} - \dfrac{c}{d} = \dfrac{a}{b} \cdot \dfrac{d}{c}$

28. $\dfrac{a}{b}$ equals $\dfrac{a+c}{b+c}$

29. $\dfrac{a}{b} + \dfrac{c}{d} = \dfrac{ad+bc}{bd}$

30. $\dfrac{a}{b} - \dfrac{c}{d} = \dfrac{ad-bc}{bd}$

31. $\dfrac{a}{b} = \dfrac{1}{b} \cdot a$

32. $\dfrac{a}{c} + \dfrac{b}{c} = \dfrac{1}{c}(a+b)$

33. $(-a)(-b) = ab$

34. $(-a)b = a(-b) = -ab$

35. $a\dfrac{b}{a} = b$

36. $\dfrac{-a}{-b} = -\dfrac{a}{b}$

37. $a + \dfrac{c}{d} = \dfrac{ad+c}{d}$

38. $a + \dfrac{b}{a} = 1 + b$

39. $a\dfrac{c}{d} = \dfrac{adc}{d}$

40. $\dfrac{3+x}{12} = \dfrac{1+x}{4}$

41. $\dfrac{1}{b} \cdot \dfrac{1}{c} = \dfrac{1}{bc}$

42. $\dfrac{1}{b} + \dfrac{1}{d} = \dfrac{1}{b+d}$

43. $a + 0 = 0$

44. $a \cdot 0 = 0$

45. $a \cdot 1 = a$

46. $a(b-c) = ab - ac$

47. $a - a = 0$

48. $a - b = a + (-b)$

49. $-(a-b) = b - a$

50. $\dfrac{a}{a} = 0$

51. $(a-b)^2 = -(b-a)^2$

52. $(a-b)^3 = -(b-a)^3$

53. $\dfrac{\frac{a}{b}}{\frac{c}{d}} = \dfrac{ad}{bc}$

54. $|x|$ is always positive

55. $|a-b| = |b-a|$

56. $\dfrac{ax+b}{ay+d} = \dfrac{x+b}{y+d}$

57. $\dfrac{a}{b-c} = \dfrac{a}{b} - \dfrac{a}{c}$

58. $\dfrac{2a}{bc} = \dfrac{a}{b} \cdot \dfrac{a}{c}$

59. $\dfrac{\frac{a}{b}}{\frac{a}{c}} = \dfrac{c}{b}$

60. $\dfrac{\frac{a}{b}}{\frac{c}{b}} = \dfrac{a}{c}$

61. $\dfrac{ab}{b} = a$

62. $a - (-b) = a + b$

63. $\dfrac{a}{ab} = b$

64. The reciprocal of x is $1/x$

65. $\dfrac{\frac{1}{a}}{\frac{b}{a}} = \dfrac{b}{a}$

66. $\dfrac{\frac{1}{a}}{\frac{1}{b}} = \dfrac{b}{a}$

67. $\dfrac{\frac{a}{1}}{b} = ab$

69. $\dfrac{\dfrac{a}{a}}{b} = b$

70. $\dfrac{\dfrac{a}{b}}{c} = \dfrac{a}{bc}$

71. $\dfrac{1}{\dfrac{1}{a}} = a$

72. $-(-a) = a$

73. $\dfrac{1}{\dfrac{1}{x} \cdot \dfrac{1}{y}} = xy$

74. $\dfrac{1}{\dfrac{1}{x} + \dfrac{1}{y}} = x + y$

75. The reciprocal of $\dfrac{2}{3}$ is $-\dfrac{2}{3}$

76. $\dfrac{\dfrac{p}{q}}{\dfrac{q}{p}} = \dfrac{p^2}{q^2}$

77. $4\dfrac{a}{b} = \dfrac{4a}{4b}$

78. $a - (b-c)^2 = a + (c-b)^2$

79. $a + bc = bc + a$

80. $a + (b+c) = c + (a+b)$

81. $\dfrac{a}{b+c} = \dfrac{a}{b} + \dfrac{a}{c}$

82. $(ab)c = (ac)(bc)$

83. $\dfrac{1}{\dfrac{a}{b} + \dfrac{c}{d}} = \dfrac{b}{a} + \dfrac{d}{c}$

84. $\dfrac{1}{\dfrac{a}{c} + \dfrac{b}{c}} = \dfrac{c}{a} + \dfrac{c}{b}$

85. $\dfrac{a}{b} + \dfrac{c}{d} = \dfrac{a+c}{bd}$

86. $\dfrac{a}{b} + \dfrac{b}{a} = 1$

87. $\dfrac{a}{c} + \dfrac{b}{c} = \dfrac{1}{c}(a+b)$

88. $\dfrac{a}{b/c} = \dfrac{ac}{b}$

89. $\dfrac{1}{a}b\dfrac{1}{c} = \dfrac{b}{ac}$

90. $\dfrac{\frac{1}{a}c}{b} = \dfrac{c}{ab}$

91. $\dfrac{a/b}{b} = \dfrac{a}{b^2}$

92. The reciprocal of b is $\dfrac{b}{1}$

93. $a(1/b + 1/c) = \dfrac{a}{b} + \dfrac{a}{c}$

94. $\dfrac{1}{\frac{1}{a} + b} = a + \dfrac{1}{b}$

95. $\dfrac{-a}{b-c} = \dfrac{a}{c-b}$

96. $\dfrac{\frac{a}{b} + c}{d} = \dfrac{a+c}{bd}$

97. $\dfrac{a-b}{c-d} = \dfrac{b-a}{d-c}$

98. $-\dfrac{b-c}{a} = \dfrac{c-b}{a}$

99. $-(-b)^2 = b^2$

100. $-\dfrac{1}{c} = c$

Drill 1B—General Principles, Signs, and Fractions, with Solutions

1. $(a+b)+c = a+(b+c)$

   True. This is the associative law of addition (2.1). Because of this law, the parentheses are not needed.

2. $a+b = b+a$

   True. See (1.1).

3. $ab = ba$

   True. See (1.1).

4. $a-b = b-a$

   False. Subtraction is not commutative.
   What is true is that $a-b = -(b-a)$. See (1.2) and (1.3).

5. $a-(b-c) = (a-b)-c$

   False. Subtraction is not associative.
   What is true is that $a-(b-c) = a-b+c$. See (2.3) and (10.9).

6. $(a/b)/c = a/(b/c)$

   False. Division is not associative. See (2.4).

7. $a/b = b/a$

   False. Division is not commutative. See (1.2) and (1.3).

8. $a(b+c) = ab+ac$

True. This is the distributive law. See (3.1).

9. $a(bc) = (ab)c$

True. This is the associative law of multiplication (2.1). Because of this law, the parentheses are not necessary.

10. $-b$ is never a positive number.

False. If b is negative, then $-b$ is positive. See law 6.

11. $\dfrac{0}{2} = 0$

True. See (4.7).

12. $3(ab) = (3a)(3b)$

False. Multiplication distributes over addition, not multiplication. See law 3.

13. $\dfrac{ac}{bc} = \dfrac{a}{b}$

True. This is the cancellation law (14).

14. $\dfrac{2}{0} = 0$

False. One cannot divide by zero. See law 7.

15. $2+3(6) = 30$

False. Multiplication comes before addition. The answer is 2+18 = 20. See law 31.

16. $\dfrac{a}{c}+\dfrac{b}{c}=\dfrac{a+b}{c}$

True. See law 16.

17. $\dfrac{a+c}{b+c}$ simplifies to $\dfrac{a}{b}$

False. This is one of the many misuses of the cancellation law. See (14.1) and (14.2).

18. $\dfrac{a+c}{b+c}=\dfrac{a+1}{b+1}$

False. Another misuse of the cancellation law.

19. $\dfrac{a+b}{ac}=\dfrac{1+b}{c}$

False. Still another cancellation law error. See (14.1) and (14.5).

20. $\dfrac{3+x}{12}$ simplifies to $\dfrac{x}{4}$

False. The cancellation law error again. It is true that $\dfrac{3x}{12}=\dfrac{x}{4}$, not the above.

21. $\dfrac{a}{c}-\dfrac{b}{c}=\dfrac{a-b}{c}$

True. The is the subtraction law for fractions (16.2).

22. $\dfrac{a}{b}\cdot\dfrac{c}{d}=\dfrac{ac}{bd}$

True. This is the multiplication law for fractions (12.1).

23. $\dfrac{a}{c}+\dfrac{b}{d}=\dfrac{a+b}{c+d}$

False. Don't confuse adding fractions with multiplying fractions. See question 22 and (12.4).

24. $\dfrac{a}{b} \div \dfrac{c}{d} = \dfrac{a}{b} \cdot \dfrac{d}{c}$

True. This is the division law for fractions (13.1).

25. $\dfrac{a}{b} - \dfrac{c}{d} = \dfrac{a-d}{b-c}$

False. Don't confuse subtracting fractions with dividing fractions. See question 24.

26. The fraction $\dfrac{a}{b}$ can be built to $\dfrac{ac}{bc}$.

True. This is the law for building fractions. See (15.1).

27. $\dfrac{a}{b} - \dfrac{c}{d} = \dfrac{a}{b} \cdot \dfrac{d}{c}$

False. Don't confuse subtracting fractions with dividing fractions. See questions 24 and 25.

28. $\dfrac{a}{b}$ equals $\dfrac{a+c}{b+c}$

False. We can multiply or divide the top and bottom of a fraction by any number but not add to or subtract from the top and bottom of a fraction. See (15.1) and (14.2).

29. $\dfrac{a}{b} + \dfrac{c}{d} = \dfrac{ad+bc}{bd}$

True. This is the addition law for fractions with unlike denominators. See (17.3).

30. $\dfrac{a}{b} - \dfrac{c}{d} = \dfrac{ad-bc}{bd}$

True. This is the subtraction law for fractions with unlike denominators. See (17.4).

31. $\dfrac{a}{b} = \dfrac{1}{b} \cdot a$

True. See (9.1).

32. $\dfrac{a}{c} + \dfrac{b}{c} = \dfrac{1}{c}(a+b)$

True. See (3.4).

33. $(-a)(-b) = ab$

True. This is the sign law for multiplication. See (10.3).

34. $(-a)b = a(-b) = -ab$

True. This is another sign law for multiplication. See (10.1).

35. $a\dfrac{b}{a} = b$

True. $a\dfrac{b}{a} = \dfrac{a}{1}\dfrac{b}{a} = \dfrac{ab}{a} = b$

36. $\dfrac{-a}{-b} = -\dfrac{a}{b}$

False. $\dfrac{-a}{-b} = \dfrac{a}{b}$. See (11.3).

37. $a + \dfrac{c}{d} = \dfrac{ad+c}{d}$

True. See (17.8).

38. $a + \dfrac{b}{a} = 1 + b$

False. Another violation of the cancellation law (14.1).

39. $a\dfrac{c}{d} = \dfrac{adc}{d}$

False. $a\dfrac{c}{d} = \dfrac{a}{1}\dfrac{c}{d} = \dfrac{ac}{d}$

40. $\dfrac{3+x}{12} = \dfrac{1+x}{4}$

False. See question 20.

41. $\dfrac{1}{b} \cdot \dfrac{1}{c} = \dfrac{1}{bc}$

True. This is a special case for the law of multiplying fractions (12.1).

42. $\dfrac{1}{b} + \dfrac{1}{d} = \dfrac{1}{b+d}$

False. See question 41. For example, if $b = 1$ and $d = 1$, the left-hand side is $\dfrac{1}{1} + \dfrac{1}{1} = 1 + 1 = 2$, while the right-hand side is $\dfrac{1}{1+1} = \dfrac{1}{2}$.

43. $a + 0 = 0$

False. $a + 0 = a$. See (4.1).

44. $a \cdot 0 = 0$

True. See (4.6).

45. $a \cdot 1 = a$

True. See (5.1).

46. $a(b-c) = ab - ac$

True. The distributive law holds for subtraction as well as addition. See (3.1).

47. $a - a = 0$

True. See (8.2).

48. $a - b = a + (-b)$

True. This is the definition of subtraction (8.1).

49. $-(a - b) = b - a$

True. $-(a - b) = -a + b = b - a$. See (10.7) and (8.3).

50. $\dfrac{a}{a} = 0$

False. $\dfrac{a}{a} = 1$. See (9.2).

51. $(a - b)^2 = -(b - a)^2$

False, because $(a - b)^2 = [-(b - a)\,]^2 = (b - a)^2$.
See (20.6) and question 49.

52. $(a - b)^3 = -(b - a)^3$

True. See (20.5) and question 49.

53. $\dfrac{\frac{a}{b}}{\frac{c}{d}} = \dfrac{ad}{bc}$

True. See (13.3).

54. $|x|$ is always positive.

False but almost true. Only false if $x = 0$.

55. $|a - b| = |b - a|$

True. Since $a - b = -(b - a), |a - b| = |-(b - a)| = |-1||b - a| = 1|b - a| = |b - a|$

56. $\dfrac{ax + b}{ay + d} = \dfrac{x + b}{y + d}$

False. Another misapplication of the cancellation law (14.1).

57. $\dfrac{a}{b-c} = \dfrac{a}{b} - \dfrac{a}{c}$

False. Try letting $a = 6$, $b = 3$, and $c = 2$. Then the left-hand side is $\dfrac{6}{3-2} = \dfrac{6}{1} = 6$, while the right-hand side is $\dfrac{6}{3} - \dfrac{6}{2} = 2 - 3 = -1$.

58. $\dfrac{2a}{bc} = \dfrac{a}{b} \cdot \dfrac{a}{c}$

False. $2a = a + a \neq a \cdot a$.

59. $\dfrac{\frac{a}{b}}{\frac{a}{c}} = \dfrac{c}{b}$

True. By (13.3) and (14.1), $\dfrac{\frac{a}{b}}{\frac{a}{c}} = \dfrac{a}{b} \dfrac{c}{a} = \dfrac{ac}{ab} = \dfrac{c}{b}$.

60. $\dfrac{\frac{a}{b}}{\frac{c}{b}} = \dfrac{a}{c}$

True. Again by (13.3) and (14.1), $\dfrac{\frac{a}{b}}{\frac{c}{b}} = \dfrac{a}{b} \dfrac{b}{c} = \dfrac{ab}{bc} = \dfrac{a}{c}$.

61. $\dfrac{ab}{b} = a$

True. By (14.1),

$$\dfrac{ab}{b} = \dfrac{a}{1} = a.$$

62. $a - (-b) = a + b$

True. See (10.1).

63. $\dfrac{a}{ab} = b$

False.

64. The reciprocal of x is $1/x$.

True. $1/x = x^{-1}$ is also called the multiplicative inverse of x. See law 7.

65. $\dfrac{\frac{1}{a}}{\frac{a}{b}} = \dfrac{b}{a}$

True. Since $\dfrac{a}{b}\cdot\dfrac{b}{a} = \dfrac{ab}{ab} = 1$, $\dfrac{a}{b}$ and $\dfrac{b}{a}$ are reciprocals of each other. See (15.6).

66. $\dfrac{\frac{1}{a}}{\frac{1}{b}} = \dfrac{b}{a}$

True. $\dfrac{\frac{1}{a}}{\frac{1}{b}} = \dfrac{1}{a}\dfrac{b}{1} = \dfrac{b}{a}.$

67. $\dfrac{\frac{a}{1}}{\frac{1}{b}} = ab$

True. $\dfrac{\frac{a}{1}}{\frac{1}{b}} = a\dfrac{b}{1} = ab$. See (9.3) and (13.3).

69. $\dfrac{\frac{a}{a}}{\frac{1}{b}} = b$

True. $\dfrac{\frac{a}{a}}{\frac{1}{b}} = a\dfrac{b}{a} = b$. Again see (9.3) and (13.3).

70. $\dfrac{\frac{a}{b}}{c} = \dfrac{a}{bc}$

True. $\dfrac{\frac{a}{b}}{c} = \dfrac{\frac{a}{b}}{\frac{c}{1}} = \dfrac{a}{b}\dfrac{1}{c} = \dfrac{a}{bc}.$

71. $\dfrac{\frac{1}{1}}{\frac{1}{a}} = a$

True. $\dfrac{\frac{1}{1}}{\frac{1}{a}} = 1\dfrac{a}{1} = a.$

The reciprocal of the reciprocal of a number is the number itself.

72. $-(-a) = a$

True. $-(-a) = (-1)(-1)(a) = (-1)^2 a = 1a = a$. See (10.6) and (20.6).

73. $\dfrac{1}{\frac{1}{x}\cdot\frac{1}{y}} = xy$

True. $\dfrac{1}{\frac{1}{x}\cdot\frac{1}{y}} = \dfrac{1}{\frac{1}{xy}} = \dfrac{xy}{1} = xy$. See question 71.

74. $\dfrac{1}{\dfrac{1}{x}+\dfrac{1}{y}} = x+y$

False. $\dfrac{1}{\dfrac{1}{x}+\dfrac{1}{y}} = \dfrac{1}{\dfrac{y}{xy}+\dfrac{x}{xy}} = \dfrac{1}{\dfrac{y+x}{xy}} = \dfrac{xy}{y+x} = \dfrac{xy}{x+y}.$

See (36.13) and compare to question 73. The reason for this common error is that $\dfrac{1}{\dfrac{1}{x}\cdot\dfrac{1}{y}}$ and $\dfrac{1}{\dfrac{1}{x}+\dfrac{1}{y}}$ look so similar.

75. The reciprocal of $\dfrac{2}{3}$ is $-\dfrac{2}{3}$.

False. The reciprocal of $\dfrac{2}{3}$ is $\dfrac{3}{2}$. The negative of $\dfrac{2}{3}$ is $-\dfrac{2}{3}$.

76. $\dfrac{\dfrac{p}{q}}{\dfrac{q}{p}} = \dfrac{p^2}{q^2}$

True. $\dfrac{\dfrac{p}{q}}{\dfrac{q}{p}} = \dfrac{p}{q}\dfrac{p}{q} = \dfrac{p^2}{q^2}.$

77. $4\dfrac{a}{b} = \dfrac{4a}{4b}$

False. $4\dfrac{a}{b} = \dfrac{4}{1}\dfrac{a}{b} = \dfrac{4a}{b}.$

78. $a-(b-c)^2 = a+(c-b)^2$

False, since $(b-c)^2 = (c-b)^2$. (See question 51.)
Then $a-(b-c)^2 = a-(c-b)^2$.

79. $a+bc = bc+a$

True. This follows from the commutative law of addition (1.1).

80. $a+(b+c) = c+(a+b)$

True. This follows from (1.1) again and is the commutative law and the associative law combined.

81. $\dfrac{a}{b+c} = \dfrac{a}{b} + \dfrac{a}{c}$

False. $\dfrac{a}{b} + \dfrac{a}{c} = \dfrac{ac+ab}{bc}$. $\dfrac{a}{b+c}$ can't be simplified.

82. $(ab)c = (ac)(bc)$

False. See question 57.

83. $\dfrac{1}{\dfrac{a}{b} + \dfrac{c}{d}} = \dfrac{b}{a} + \dfrac{d}{c}$

False. If the *plus* were *times*, it would be true.

84. $\dfrac{1}{\dfrac{a}{c} + \dfrac{b}{c}} = \dfrac{c}{a} + \dfrac{c}{b}$

False. $\dfrac{1}{\dfrac{a}{c} + \dfrac{b}{c}} = \dfrac{1}{\dfrac{a+b}{c}} = \dfrac{c}{a+b}.$

85. $\dfrac{a}{b} + \dfrac{c}{d} = \dfrac{a+c}{bd}$

False. See (17.3).

86. $\dfrac{a}{b} + \dfrac{b}{a} = 1$

False. If the *plus* were *times*, it would be true.

87. $\dfrac{a}{c} + \dfrac{b}{c} = \dfrac{1}{c}(a+b)$

True. See (3.4).

88. $\dfrac{a}{\dfrac{b}{c}} = \dfrac{b}{c}$

False. $\dfrac{a}{\dfrac{b}{c}} = a\dfrac{c}{b} = \dfrac{ac}{b}.$

89. $\dfrac{1}{a}b\dfrac{1}{c} = \dfrac{b}{ac}$

True. $\dfrac{1}{a}b\dfrac{1}{c} = \dfrac{1}{a}\dfrac{b}{1}\dfrac{1}{c} = \dfrac{b}{ac}$.

90. $\dfrac{\frac{1}{a}c}{b} = \dfrac{c}{ab}$

True. $\dfrac{\frac{1}{a}c}{b} = \dfrac{\frac{c}{a}}{b} = \dfrac{c}{a}\dfrac{1}{b} = \dfrac{c}{ab}$.

91. $\dfrac{\frac{a}{b}}{b} = a$

False. $\dfrac{\frac{a}{b}}{b} = \dfrac{a}{b}\dfrac{1}{b} = \dfrac{a}{b^2}$.

92. The reciprocal of b is $\dfrac{b}{1}$.

False. It is $\dfrac{1}{b}$. See law 7.

93. $a(1/b + 1/c) = \dfrac{a}{b} + \dfrac{a}{c}$

True. $a(1/b + 1/c) = a/b + a/c = \dfrac{a}{b} + \dfrac{a}{c}$.

94. $\dfrac{1}{\frac{1}{a} + b} = a + \dfrac{1}{b}$

False. If the *plus* were *times,* it would be true. We can simplify by multiplying top and bottom by a,

$$\dfrac{1}{\frac{1}{a}+b} = \dfrac{1}{\frac{1}{a}+b}\dfrac{a}{a} = \dfrac{a}{\frac{1}{a}a+ba} = \dfrac{a}{1+ba} = \dfrac{a}{1+ab}.$$

95. $\dfrac{-a}{b-c} = \dfrac{a}{c-b}$

True. Multiply top and bottom by -1. $\dfrac{-a(-1)}{(b-c)(-1)} = \dfrac{a}{-b+c} = \dfrac{a}{c-b}$.

96. $\dfrac{\dfrac{a}{b}+c}{d}=\dfrac{a+c}{bd}$

False. Multiply top and bottom by b: $\dfrac{\dfrac{a}{b}+c}{d}\cdot\dfrac{b}{b}=\dfrac{\dfrac{a}{b}b+cb}{bd}=\dfrac{a+cb}{bd}=\dfrac{a+bc}{bd}$.

97. $\dfrac{a-b}{c-d}=\dfrac{b-a}{d-c}$

True. Multiply top and bottom by -1: $\dfrac{-1(a-b)}{-1(c-d)}=\dfrac{-a+b}{-c+d}=\dfrac{b-a}{d-c}$.

98. $-\dfrac{b-c}{a}=\dfrac{c-b}{a}$

True. $-\dfrac{b-c}{a}=\dfrac{-(b-c)}{a}=\dfrac{-b+c}{a}=\dfrac{c-b}{a}$.

99. $-(-b)^2=b^2$

False, since $(-b)^2=b^2$, $-(-b)^2=-(b^2)=-b^2$.

100. $-\dfrac{1}{c}=c$

False. Negative and reciprocal don't cancel. See laws 6 and 7. $-\dfrac{1}{c}$ doesn't simplify.

Drill 2A—Exponents and Roots

True or false. (50 questions)

1. $x^3 x^4 = x^7$

2. $\dfrac{x^5}{x^2} = x^3$

3. $x^2 y^2 = (xy)^2$

4. $\left(\dfrac{x}{y}\right)^3 = \dfrac{x^3}{y^3}$

5. $(x^2)^3 = x^6$

6. $x^{-3} = \dfrac{1}{x^3}$

7. $\dfrac{x^5}{y^2} = \left(\dfrac{x}{y}\right)^3$

8. $\dfrac{x^7}{x^{-2}} = x^5$

9. $x^{-2} = \sqrt{x}$

10. $x^{-1/3} = \dfrac{1}{\sqrt[3]{x}}$

11. $-3^2 = 9$

12. $\sqrt{9} = \pm 3$

13. $(x^3)^4 = x^7$

14. $\sqrt{x^2}$ is always equal to x

15. $\sqrt[3]{x^3}$ is always equal to x

16. $(x^{-1} y^{-1})^{-1} = xy$

17. $b^2 b^{-2} = 1$

18. $(a+b)^{-1} = a^{-1} + b^{-1}$

19. $x^2 + x^{-2} = 1$

20. $3x^2$ means 9 times x^2

21. $xx^{-1} = 0$

22. $a^2 a^3 = a^6$

23. $(3+x)^2 = 9 + x^2$

24. $\sqrt[3]{x^3 y} = x\sqrt[3]{y}$

25. $y^2 + y^3 = y^5$

26. $\sqrt{b^2 + c^2} = b + c$

27. $\sqrt{x^2 y} = x\sqrt{y}$

28. $a^2 - b^2 = a^2 - 2ab + b^2$

29. $-x^2$ is never negative

30. $\sqrt[3]{a^3 + b} = a + \sqrt[3]{b}$

31. $\sqrt{x}\sqrt{y} = \sqrt{xy}$

32. $\sqrt{5} + \sqrt{2} = \sqrt{10}$

33. $(\sqrt{x} + \sqrt{y})^2 = x + y$

34. $3^{-2} = -9$

35. $(-x)^0 = -1$

36. x^2 means two times x

37. $\sqrt[3]{(x+y)^3} = x+y$

38. $5(y+y+y) = 5y^3$

39. $\dfrac{10^6}{5^6} = 2^6$

40. $\sqrt{(x+y)^2} = x+y$

41. $\dfrac{\sqrt{x}}{\frac{1}{\sqrt{x}}} = x$

42. $a^{-1} + b^{-1} = \dfrac{1}{a+b}$

43. $(x-y)^2 = (y-x)^2$

44. $(x-y)^3 = (y-x)^3$

45. $\left(\dfrac{b}{a}\right)^{-1} = \dfrac{a}{b}$

46. $\left(\dfrac{1}{x}\right)^{-1} = x$

47. $2x^{-1} = \dfrac{1}{2x}$

48. $x^{1/2} x^{1/3} = x^{1/6}$

49. $\left(\dfrac{y}{x}\right)^{-n} = \left(\dfrac{x}{y}\right)^{n}$

50. $\dfrac{\sqrt{x}}{x} = \dfrac{1}{\sqrt{x}}$

Drill 2B—Exponents and Roots, with Solutions

True or false. (50 questions)

1. $x^3 x^4 = x^7$

   True. See (23.1).

2. $\dfrac{x^5}{x^2} = x^3$

   True. See (23.2).

3. $x^2 y^2 = (xy)^2$

   True. See (24.1).

4. $\left(\dfrac{x}{y}\right)^3 = \dfrac{x^3}{y^3}$

   True. See (24.2).

5. $(x^2)^3 = x^6$

   True. See (25.1).

6. $x^{-3} = \dfrac{1}{x^3}$

   True. See (20.1).

7. $\dfrac{x^5}{y^2} = \left(\dfrac{x}{y}\right)^3$

   False. Neither the exponents nor the bases are the same, so neither (23.2) nor (24.2) applies.

8. $\dfrac{x^7}{x^{-2}} = x^5$

   False. $\dfrac{x^7}{x^{-2}} = x^{7-(-2)} = x^{7+2} = x^9$.

9. $x^{-2} = \sqrt{x}$

   False. $x^{-2} = \dfrac{1}{x^2}$ but $\sqrt{x} = x^{1/2}$.

10. $x^{-1/3} = \dfrac{1}{\sqrt[3]{x}}$

    True. $x^{-1/3} = \dfrac{1}{x^{1/3}} = \dfrac{1}{\sqrt[3]{x}}$.

11. $-3^2 = 9$

    False. $-3^2 = -9$. The minus is not squared only the three. Powers are done before negation.

12. $\sqrt{9} = \pm 3$

    False. $\sqrt{9} = +3$ only. $\sqrt{x}$ means the *principal* or *positive* square root of x. See law 21.

13. $(x^3)^4 = x^7$

    False. $(x^3)^4 = x^{3 \cdot 4} = x^{12}$. See (25.1).

14. $\sqrt{x^2}$ is always equal to x.

    False. It is true only if x is positive or zero but not negative. See (33.6).

15. $\sqrt[3]{x^3}$ is always equal to x.

    True. See (33.7).

16. $(x^{-1}y^{-1})^{-1} = xy$

    True. $(x^{-1}y^{-1})^{-1} = (x^{-1})^{-1}(y^{-1})^{-1} = x^1 y^1 = xy$.

17. $b^2 b^{-2} = 1$

True. $b^2 b^{-2} = b^{2-2} = b^0 = 1$.

18. $(a+b)^{-1} = a^{-1} + b^{-1}$

False. $(ab)^{-1} = a^{-1} b^{-1}$. True for *times* but not *plus*.

19. $x^2 + x^{-2} = 1$

False. True if the plus sign were a times sign. See question 17.

20. $3x^2$ means 9 times x^2

False. The 3 is not squared, only the x.

21. $xx^{-1} = 0$

False. $xx^{-1} = x^{1-1} = x^0 = 1$.

22. $a^2 a^3 = a^6$

False. The exponents should be added not multiplied.
$a^2 a^3 = a^{2+3} = a^5$.

23. $(3+x)^2 = 9 + x^2$

False. True if the plus signs were times. $(3x)^2 = 9x^2$.

24. $\sqrt[3]{x^3 y} = x\sqrt[3]{y}$

True. $\sqrt[3]{x^3 y} = \sqrt[3]{x^3} \sqrt[3]{y} = x\sqrt[3]{y}$.

25. $y^2 + y^3 = y^5$

False. True for times but not plus.
$y^2 y^3 = y^5$, but $y^2 + y^3$ cannot be simplified.

26. $\sqrt{b^2 + c^2} = b + c$

False. True for times but not plus. $\sqrt{b^2 c^2} = |bc|$.

27. $\sqrt{x^2 y} = x\sqrt{y}$

False, but true if x is positive or zero. $\sqrt{x^2 y} = \sqrt{x^2}\sqrt{y} = |x|\sqrt{y}$.

28. $a^2 - b^2 = a^2 - 2ab + b^2$

False. $a^2 - b^2 = (a+b)(a-b)$, but $a^2 - 2ab + b^2 = (a-b)^2$.

29. $-x^2$ is never negative.

False. It is always negative or zero. It is the negative of x^2 not $(-x)^2$.

30. $\sqrt[3]{a^3 + b} = a + \sqrt[3]{b}$

False. True for times but not plus. See (33.9), (33.10), and question 26.

31. $\sqrt{x}\sqrt{y} = \sqrt{xy}$

True. See (34.1). x and y must be positive or zero.

32. $\sqrt{5} + \sqrt{2} = \sqrt{10}$

False. True for times but not plus. See (34.1) and question 31.

33. $(\sqrt{x} + \sqrt{y})^2 = x + y$

False. True if the plus were times.
$(\sqrt{x}\sqrt{y})^2 = (\sqrt{x})^2(\sqrt{y})^2 = xy$

$(\sqrt{x} + \sqrt{y})^2 = x + 2\sqrt{x}\sqrt{y} + y$.

34. $3^{-2} = -9$

False. $3^{-2} = \dfrac{1}{3^2} = \dfrac{1}{9}.$

35. $(-x)^0 = -1$

False. Any number (except 0), negative or positive, to the zero power is one. $(-x)^0 = +1.$

36. x^2 means two times x

False. x^2 means "x times x," not "2 times x."

37. $\sqrt[3]{(x+y)^3} = x+y$

True. See (33.7).

38. $5(y+y+y) = 5y^3$

False. $5(yyy) = 5y^3$, but $5(y+y+y) = 5(3y) = 15y.$

39. $\dfrac{10^6}{5^6} = 2^6$

True. $\dfrac{10^6}{5^6} = \left(\dfrac{10}{5}\right)^6 = 2^6.$

40. $\sqrt{(x+y)^2} = x+y$

False, but true if $x+y$ is positive. $\sqrt{(x+y)^2} = |x+y|.$

41. $\dfrac{\sqrt{x}}{\dfrac{1}{\sqrt{x}}} = x$

True. $\dfrac{\sqrt{x}}{\dfrac{1}{\sqrt{x}}} = \dfrac{\sqrt{x}}{1}\dfrac{\sqrt{x}}{1} = \sqrt{x}\sqrt{x} = x.$

Of course, x must be positive or zero.

42. $a^{-1} + b^{-1} = \dfrac{1}{a+b}$

False. $a^{-1} + b^{-1} = \dfrac{1}{a} + \dfrac{1}{b} \neq \dfrac{1}{a+b}.$

43. $(x - y)^2 = (y - x)^2$

True. See question 51 of Drill 1.

44. $(x - y)^3 = (y - x)^3$

False. Since $(x - y) = -(y - x)$, $(x - y)^3 = -(y - x)^3$

45. $\left(\dfrac{b}{a}\right)^{-1} = \dfrac{a}{b}$

True. $\left(\dfrac{b}{a}\right)^{-1} = \dfrac{b^{-1}}{a^{-1}} = \dfrac{1/b}{1/a} = \dfrac{1}{b} \cdot \dfrac{a}{1} = \dfrac{a}{b}$.

46. $\left(\dfrac{1}{x}\right)^{-1} = x$

True. $\left(\dfrac{1}{x}\right)^{-1} = (x^{-1})^{-1} = x^1 = x$.

47. $2x^{-1} = \dfrac{1}{2x}$

False. $2x^{-1} = 2\dfrac{1}{x} = \dfrac{2}{x}$.

48. $x^{1/2}x^{1/3} = x^{1/6}$

False. $x^{1/2}x^{1/3} = x^{1/2+1/3} = x^{5/6}$.

49. $\left(\dfrac{y}{x}\right)^{-n} = \left(\dfrac{x}{y}\right)^{n}$

True. $\left(\dfrac{y}{x}\right)^{-n} = \dfrac{y^{-n}}{x^{-n}} = \dfrac{\left(\dfrac{1}{y^n}\right)}{\left(\dfrac{1}{x^n}\right)} = \dfrac{1}{y^n} \cdot \dfrac{x^n}{1} = \dfrac{x^n}{y^n} = \left(\dfrac{x}{y}\right)^{n}$.

See question 45. Note that inverting a fraction is equivalent to changing the sign of the exponent.

50. $\dfrac{\sqrt{x}}{x} = \dfrac{1}{\sqrt{x}}$

True. Multiply top and bottom by $\sqrt{x}$: $\dfrac{\sqrt{x}}{x} \cdot \dfrac{\sqrt{x}}{\sqrt{x}} = \dfrac{x}{x\sqrt{x}} = \dfrac{1}{\sqrt{x}}$.

Drill 3A—Linear Equations

True or False. (50 questions)

1. If $a = b$, then $b = a$.

2. If If $a = b$ and $b = c$, then $a = c$.

3. If $a = b$, then $a + c = b + c$.

4. If $a = b$, then $ac = bc$.

5. If $a - b = 0$, then $a = b$.

6. If $\dfrac{a}{b} = 1$, then $a = b$.

7. If $ab = 1$, then $b = 1/a$.

8. If $\dfrac{a}{b} = \dfrac{c}{d}$, then $ad = bc$.

9. If $a = b$, then $b = -a$.

10. If $a + b = 0$, then $b = 1/a$.

11. If $ab = 0$, then $a = 0$ or $b = 0$.

12. If $a/b = 0$, then $a = 0$ and $b \neq 0$.

13. If x=y, then $x/c = y/c$.

14. If $a + b = 0$, then $a = 0$ or $b = 0$.

15. If $ab = 1$, then $b = -a$.

16. $\dfrac{a = b}{c}$ means $\dfrac{a}{c} = \dfrac{b}{c}$.

106

17. If $a/b = 0$, then $b = 0$.

18. If $a = 1/b$, then $b = -a$.

19. If $y = -x$, then $x = -y$.

20. If $6 + x = 2$, then $x = 4$.

21. If 6x=3, then x = 2.

22. If $\dfrac{x}{3}$=6, then x=2.

23. If $\dfrac{x}{2} = \dfrac{1}{3}$, then $x = \dfrac{2}{3}$.

24. If $2/a = 10$, then $a = 5$.

25. If $\dfrac{x/6}{3} = 1$, then $x = 2$.

26. If $\dfrac{a}{b} = \dfrac{3}{4}$, then $a = 3$ and $b = 4$.

27. If $\dfrac{x}{2} = \dfrac{3}{a}$, then $x = \dfrac{6}{a}$.

28. If $\dfrac{3}{x} = \dfrac{1}{b}$, then $x = \dfrac{b}{3}$.

29. If $3x = 4$, then $x = 3/4$.

30. If $\dfrac{x}{2} = \dfrac{y}{2} + 1$, then $x = y + 1$.

31. If $x = 3 - y$, then $y = 3 - x$.

32. If $(x-1)(x+5)$=0, then $x = 1$ or $x = -5$.

33. If $2x + y = 6$, then $x + y = 3$.

34. If $3 + \dfrac{x}{2}$=a, then $x = 2a + 3$.

35. If $x+2=10$, then $x=5$.

36. If 6-x=10, then $x=4$.

37. If $2x=12$, then $x=10$.

38. If $\dfrac{1}{x}-3=1$, then $x=4$.

39. If $\dfrac{2}{x}=3$, then $x=2/3$.

40. If $(x-1)(x-3)=5$, then $x=6$ or $x=8$.

41. If $\dfrac{x}{2}+1=x-3$, then $x+1=2x-6$.

42. If $\dfrac{x}{y}=\dfrac{2}{3}$, then $y=\dfrac{2}{3}x$.

43. The only solution of $|x+2|=10$ is $x=8$.

44. If $\dfrac{1}{2}(x-3)=4$, then $x-6=8$.

45. If 3+2x=7, then $x=7/5$.

46. If 5(x-10)=0, then $x=2$.

47. If x(x-1)(x-4)=0, then $x=0, 1$, or 4.

48. If $\dfrac{1}{x}=\dfrac{1}{a}+\dfrac{1}{b}$, then $x=a+b$.

49. If $\dfrac{x}{c}+$b=a, then $x=ac-b$.

50. If b+cx=a, then $x=\dfrac{a}{b+c}$.

Drill 3B—Linear Equations, with Solutions

True or False. (50 Questions)

1. If $a = b$, then $b = a$.

    True. See (26.2).

2. If $a = b$, and $b = c$, then $a = c$.

    True. See (26.3).

3. If $a = b$, then $a + c = b + c$.

    True. See (28.1).

4. If $a = b$, then $ac = bc$.

    True. See (28.3).

5. If $a - b = 0$, then $a = b$.

    True. Add b to both sides.

6. If $\dfrac{a}{b} = 1$, then $a = b$.

    True. Multiply both sides by b.

7. If $ab = 1$, then $b = 1/a$.

    True. Divide both sides by a.

8. If $\dfrac{a}{b} = \dfrac{c}{d}$, then $ac = bd$.

    False. $ad = bc$. Or multiply both sides by bd and cancel as follows:
    $\dfrac{a}{b}bd = \dfrac{c}{d}bd$, $\dfrac{a}{1}d = \dfrac{c}{1}b$, $ad = bc$.

See also (28.5), cross multiplication.

9. If $a = b$, then $b = -a$.

    False. If $a = b$ then $b = a$. See question 1.

10. If $a + b = 0$, then $b = 1/a$.

    False. $b = -a$. Subtract a from both sides.

11. If $ab = 0$, then $a = 0$ or $b = 0$.

    True. See (39.6).

12. If $\dfrac{a}{b} = 0$, then $a = 0$ and $b \neq 0$.

    True. First, $b \neq 0$, since if $b=0$, then a/b would not exist. See (7.3).
    Second, $a = 0$, since we can multiply both sides by b and

    $$\dfrac{a}{b}b = b \cdot 0 = 0$$

    $a = 0$.

13. If $x=y$, then $x/c = y/c$.

    True. We can divide both sides by c. $c \neq 0$ of course.

14. If $a + b = 0$, then $a = 0$ or $b = 0$.

    False. If the plus sign were times, it would be true.
    If $a + b = 0$ then $b = -a$.

110

15. If $ab = 1$, then $b = -a$.

False. See question 7.

16. $\dfrac{a = b}{c}$ means $\dfrac{a}{c} = \dfrac{b}{c}$.

False. $\dfrac{a = b}{c}$ is meaningless. You can't divide an equation by a number. You can, however, divide each side by a number.

17. If $a/b = 0$, then $b = 0$.

False. See question 12.

18. If $a = 1/b$, then $b = -a$.

False. If $a = 1/b$ then $ab = 1$ and $b = 1/a$.

19. If $y = -x$, then $x = -y$.

True. If $y = -x$ then multiply both sides by -1 to get $-y = x$ or $x = -y$.

20. If $6 + x = 2$, then $x = 4$.

False. Subtracting 6 from both sides gives $x = 2 - 6 = -4$.

21. If $6x = 3$, then $x = 2$.

False. Dividing both sides by 6 gives $x = \dfrac{3}{6} = \dfrac{1}{2}$.

22. If $\dfrac{x}{3} = 6$, then $x = 2$.

False. Multiplying both sides by 3 gives $x = 3 \cdot 6 = 18$.

23. If $\dfrac{x}{2} = \dfrac{1}{3}$, then $x = \dfrac{2}{3}$.

True. Multiply both sides by 2. $\dfrac{x}{2} \cdot 2 = 2 \cdot \dfrac{1}{3}$ or $x = \dfrac{2}{3}$.

24. If $2/a = 10,$ then $a = 5.$

False. Multiply both sides by a, $2 = 10a.$

Then divide both sides by 10, $\dfrac{2}{10} = a$ or $a = \dfrac{1}{5}.$

25. If $\dfrac{x/6}{3} = 1,$ then $x = 2.$

False. Multiply both sides by 3, $\dfrac{x}{6} = 3.$

The multiply both sides by 6, $6\dfrac{x}{6} = 6 \cdot 3$ or $x = 18.$

26. If $\dfrac{a}{b} = \dfrac{3}{4},$ then $a = 3$ and $b = 4.$

False. Suppose $a = 6$ and $b = 8$. Then $\dfrac{a}{b} = \dfrac{6}{8} = \dfrac{3}{4}.$

27. If $\dfrac{x}{2} = \dfrac{3}{a},$ then $x = \dfrac{6}{a}.$

True. Multiply both sides by 2, $2\dfrac{x}{2} = 2\dfrac{3}{a}$ or $x = \dfrac{2 \cdot 3}{a} = \dfrac{6}{a}.$

28. If $\dfrac{3}{x} = \dfrac{1}{b},$ then $x = \dfrac{b}{3}.$

False. If $\dfrac{3}{x} = \dfrac{1}{b}$ then cross multipy to get $x = 3b.$

29. If $3x = 4,$ then $x = 3/4.$

False. Divide both sides by 3. If $\dfrac{3x}{3} = \dfrac{4}{3}$ then $x = \dfrac{4}{3} = 1\dfrac{1}{3}.$

30. If $\dfrac{x}{2} = \dfrac{y}{2} + 1,$ then $x = y + 1.$

False. Multiply both sides by 2, $2\dfrac{x}{2} = 2(\dfrac{y}{2} + 1)$. Then $x = 2\dfrac{y}{2} + 2 \cdot 1 = y + 2.$

31. If $x = 3 - y,$ then $y = 3 - x.$

True. Add y to both sides and then subtract x from both sides.
$x = 3 - y,\ \ x + y = 3,\ \ y = 3 - x.$

32. If $(x-1)(x+5)=0$, then $x = 1$ or $x = -5$.

True. According to (39.6), $x - 1 = 0$ or $x + 5 = 0$, so $x = 1$ or $x = -5$.

33. If $2x + y = 6$, then $x + y = 3$.

False. Dividing both sides by 2 gives $\dfrac{2x+y}{2} = \dfrac{6}{2}$

$x + \dfrac{y}{2} = 3$.

34. If $3 + \dfrac{x}{2} = a$, then $x = 2a - 3$.

False. Multiply both sides by 2, $2\left(3 + \dfrac{x}{2}\right) = 2a$, $6 + 2\dfrac{x}{2} = 2a$, $6 + x = 2a$.

Subtract 6 from both sides, x =2a-6.

35. If $x + 2 = 10$, then $x = 5$.

False. Subtract 2 from both sides. $x = 10 - 2 = 8$.

36. If $6 - x = 10$, then $x = 4$.

False. Subtract 6 from both sides, $6 - x - 6 = 10 - 6$, $-x = 4$, $x = -4$.

37. If $2x = 12$, then $x = 10$.

False. Divide both sides by 2. $\dfrac{2x}{2} = \dfrac{10}{2}$, $x = 5$.

38. If $\dfrac{1}{x} - 3 = 1$, then $x = 4$

False. Add 3 to both sides. $\dfrac{1}{x} - 3 + 3 = 1 + 3$, $\dfrac{1}{x} = 4$.

Now multiply both sides by x and divide both sides by 4:

$1 = 4x$, $\dfrac{1}{4} = x$, $x = \dfrac{1}{4}$.

39. If $\dfrac{2}{x} = 3$, then $x = 2/3$.

True. Multiply both sides by x, and then divide both sides by 3.

$$x\frac{2}{x} = 3x \quad 2 = 3x \quad \frac{2}{3} = x \quad x = \frac{2}{3}.$$

40. If $(x-1)(x-3)=5$, then $x = 6$ or $x = 8$.

False. This is a misapplication of (39.5). The right side must be zero, nothing else. This equation is a quadratic equation and may be solved by the quadratic formula.

Multiply the left side and collect like terms.
$$(x-1)(x-3) = x^2 - x - 3x + 3 = x^2 - 4x + 3 = 5$$
or $x^2 - 4x + 3 - 5 = 0$, $x^2 - 4x - 2 = 0$.

Now using the quadratic formula,

$$x = \frac{-(-4) \pm \sqrt{(-4)^2 - 4(1)(-2)}}{2(1)} = \frac{4 \pm \sqrt{16+8}}{2} = \frac{4 \pm \sqrt{24}}{2}$$

$$x = \frac{4 \pm \sqrt{4 \cdot 6}}{2} = \frac{4 \pm 2\sqrt{6}}{2} = \frac{2(2 \pm \sqrt{6})}{2} = 2 \pm \sqrt{6}.$$

41. If $\dfrac{x}{2} + 1 = x - 3$, then $x + 1 = 2x - 6$.

False. Multiply both sides by 2: $2\left(\dfrac{x}{2} + 1\right) = 2(x-3)$ $x + 2 = 2x - 6$. The left side is $x + 2$, not $x + 1$.

42. If $\dfrac{x}{y} = \dfrac{2}{3}$, then $y = \dfrac{2}{3}x$.

False. Use (28.5) to cross multiply: $3x = 2y$.

Then divide both sides by 2: $\dfrac{3x}{2} = y$ or $y = \dfrac{3}{2}x$.

43. The only solution of $|x+2|=10$ is $x = 8$.

False. Another solution is –12.
If $|x+2|=10$, then by (18.3), $x+2 = \pm 10$ or $x = -2 \pm 10 = -12$ or 8.

44. If $\frac{1}{2}(x-3)=4$, then $x - 6 = 8$.

False. Multiply both sides by 2, $2\frac{1}{2}(x-3)=2 \cdot 4$ or $x - 3 = 8$.
The left side is $x - 3$ not $x - 6$.

45. If $3+2x=7$, then $x = 7/5$.

False. Subtract 3 from both sides and then divide both sides by 2.
$3 + 2x - 3 = 7 - 3$, $2x = 4$, $x = 2$.

46. If $5(x-10)=0$, then $x = 2$.

False. Divide both sides by 5: $\dfrac{5(x-10)}{5} = \dfrac{0}{5} = 0$, $x - 10 = 0$, $x = 10$.

47. If $x(x-1)(x-4)=0$, then $x = 0$, 1, or 4.

True. See (39.6) and question 31.

48. If $\frac{1}{x} = \frac{1}{a} + \frac{1}{b}$, then $x = a + b$.

False. Multiply both sides by the least common denominator, which is abx.

$abx\frac{1}{x} = abx\frac{1}{a} + abx\frac{1}{b}$. Now cancel: $ab = bx + ax = x(b+a)$.

Now divide both sides by $b + a = a + b$, $x = \dfrac{ab}{a+b}$.

49. If $\dfrac{x}{c} + b = a$, then $x = ac - b$.

False. Multiply both sides by c:

$$c\left(\dfrac{x}{c} + b\right) = ac, \; x + cb = ac, \; x = ac - cb = ac - bc.$$

50. If $b + cx = a$, then $x = \dfrac{a}{b+c}$.

False. Subtract b from both sides and then divide both sides by c:

$$cx = a - b, \; x = \dfrac{a-b}{c}.$$

Drill 4A—Literal Equations

Multiple choice. Solve the given equation for x and choose the correct answer. (20 questions)

1. $a + x = b$

   a) a/b b) $-ab$ c) b/a d) $a-b$ e) $b-a$

2. $a - x = b$

   a) $b-a$ b) $-a/b$ c) a/b d) $a-b$ e) b/a

3. $ax = b$

   a) $b = a$ b) $-b/a$ c) $a-b$ d) ab e) b/a

4. $\dfrac{x}{a} = b$

   a) $a-b$ b) b/a c) ab d) $b-a$ e) $a-b$

5. $\dfrac{x}{a} = \dfrac{1}{b}$

   a) $a-b$ b) b/a c) ab d) a/b e) $b-a$

6. $\dfrac{a}{x} = b$

   a) a/b b) b/a c) ab d) $b-a$ e) $a-b$

7. $b - ax = c$

   a) $\dfrac{b-c}{a}$ b) $\dfrac{c-b}{a}$ c) $ac-b$ d) $\dfrac{c+b}{a}$ e) $b-c+a$

8. $a - bx = c - dx$

   a) $\dfrac{a+c}{b+d}$ b) $\dfrac{abd}{2bd}$ c) $\dfrac{a-b+c-d}{2}$ d) $\dfrac{c-a}{b+d}$ e) $\dfrac{c+a}{b-d}$

9. $\dfrac{a}{x} + b = c$

   a) $\dfrac{b-c}{a}$ b) $\dfrac{c-b}{a}$ c) $ac - b$ d) $\dfrac{b+c}{a}$ e) $\dfrac{a}{c-b}$

10. $\dfrac{x/a}{b} = c$

   a) $\dfrac{b-c}{a}$ b) $\dfrac{bc}{a}$ c) abc d) $\dfrac{c+b}{a}$ e) $\dfrac{c}{ab}$

11. $a = b(c - dx)$

   a) $\dfrac{a}{b(c-d)}$ b) $\dfrac{a-bc}{d}$ c) $ac - bd$ d) $\dfrac{b+c}{ad}$ e) $\dfrac{bc-a}{bd}$

12. $\dfrac{a}{b/x} = c$

   a) $\dfrac{b-c}{a}$ b) $\dfrac{b+c}{a}$ c) $\dfrac{ac}{b}$ d) $\dfrac{bc}{a}$ e) abc

13. $a\left(b + \dfrac{c}{x}\right) = d$

   a) $\dfrac{a-b}{cd}$ b) $\dfrac{ac}{d-ab}$ c) $\dfrac{ac}{ab-d}$ d) $\dfrac{d-ab}{c}$ e) $\dfrac{ab-d}{c}$

14. $a = b \cdot \dfrac{x}{x-b}$

   a) $\dfrac{a}{b(1-b)}$ b) $\dfrac{-a}{b}$ c) $ac - b$ d) $\dfrac{ab}{a-b}$ e) $b\text{-}a$

15. $\dfrac{1}{a} + \dfrac{1}{x} = \dfrac{1}{b}$

   a) $\dfrac{a-b}{ab}$ b) $\dfrac{ab}{a-b}$ c) $a - b$ d) $\dfrac{b-a}{a}$ e) $\dfrac{b-a}{b}$

16. $\dfrac{x-a}{x^2 + a^2} = 0$

   a) a^3 b) $\dfrac{1-a}{a^2}$ c) $a(x^2 + a^2)$ d) a e) no solution

17. $ax + x = b$

   a) $\dfrac{b}{a}$ b) $\dfrac{a+1}{b}$ c) $\dfrac{b-x}{a}$ d) $\dfrac{b}{a+1}$ e) $b(a+1)$

18. $\dfrac{x-a}{x} = b$

   a) $\dfrac{a}{1-b}$ b) $\dfrac{1-b}{a}$ c) $a + bx$ d) a e) $a + b$

19. $\dfrac{a}{b + \dfrac{c}{x}} = d$

   a) $\dfrac{cd}{a - bd}$ b) $\dfrac{c}{ad - b}$ c) $\dfrac{(b+c)d}{a}$ d) $d - a - b - c$ e) $\dfrac{a}{d(b+c)}$

20. $\dfrac{a - bx}{c - x} = d$

   a) $\dfrac{a}{cd(b-1)}$ b) $a - cd - b - 1$ c) $\dfrac{(b+c)d}{a}$ d) $\dfrac{a - cd}{b - d}$ e) $\dfrac{ab}{cd}$

Drill 4B—Literal Equations, with Solutions

Multiple choice. Solve for x. (20 questions)

1. $a + x = b$

    a) a/b b) $-ab$ c) b/a d) $a\text{-}b$ e) $b\text{-}a$

    Answer: e).

2. $a - x = b$

    a) $b\text{-}a$ b) $-a/b$ c) a/b d) $a\text{-}b$ e) b/a

    Answer: d).

3. $ax = b$

    a) $b = a$ b) $-b/a$ c) $a\text{-}b$ d) ab e) b/a

    Answer: e).

4. $\dfrac{x}{a} = b$

    a) $a\text{-}b$ b) b/a c) ab d) $b\text{-}a$ e) $a\text{-}b$

    Answer: c).

5. $\dfrac{x}{a} = \dfrac{1}{b}$

    a) $a\text{-}b$ b) b/a c) ab d) a/b e) $b\text{-}a$

    Answer: d).

6. $\dfrac{a}{x} = b$

   a) a/b b) b/a c) ab d) $b-a$ e) $a-b$

   Answer: a).

7. $b - ax = c$

   a) $\dfrac{b-c}{a}$ b) $\dfrac{c-b}{a}$ c) $ac-b$ d) $\dfrac{c+b}{a}$ e) $b-c+a$

   Answer: a).

8. $a+bx = c-dx$

   a) $\dfrac{a+c}{b+d}$ b) $\dfrac{abd}{2bd}$ c) $\dfrac{a-b+c-d}{2}$ d) $\dfrac{c-a}{b+d}$ e) $\dfrac{c+a}{b-d}$

   Answer: d).

9. $\dfrac{a}{x} + b = c$

   a) $\dfrac{b-c}{a}$ b) $\dfrac{c-b}{a}$ c) $ac-b$ d) $\dfrac{b+c}{a}$ e) $\dfrac{a}{c-b}$

   Answer: e).

10. $\dfrac{x/a}{b} = c$

   a) $\dfrac{b-c}{a}$ b) $\dfrac{bc}{a}$ c) abc d) $\dfrac{c+b}{a}$ e) $\dfrac{c}{ab}$

   Answer: c).

11. $a = b(c - dx)$

   a) $\dfrac{a}{b(c-d)}$ b) $\dfrac{a-bc}{d}$ c) $ac-bd$ d) $\dfrac{b+c}{ad}$ e) $\dfrac{bc-a}{bd}$

   Answer: e).

12. $\dfrac{a}{b/x} = c$

    a) $\dfrac{b-c}{a}$ b) $\dfrac{b+c}{a}$ c) $\dfrac{ac}{b}$ d) $\dfrac{bc}{a}$ e) abc

    Answer: d).

13. $a(b+\dfrac{c}{x}) = d$

    a) $\dfrac{a-b}{cd}$ b) $\dfrac{ac}{d-ab}$ c) $\dfrac{ac}{ab-d}$ d) $\dfrac{d-ab}{c}$ e) $\dfrac{ab-d}{c}$

    Answer: b).

14. $a = b \cdot \dfrac{x}{x-b}$

    a) $\dfrac{a}{b(1-b)}$ b) $\dfrac{-a}{b}$ c) $ac-b$ d) $\dfrac{ab}{a-b}$ e) $b\text{-}a$

    Answer: d).

15. $\dfrac{1}{a}+\dfrac{1}{x}=\dfrac{1}{b}$

    a) $\dfrac{a-b}{ab}$ b) $\dfrac{ab}{a-b}$ c) $a-b$ d) $\dfrac{b-a}{a}$ e) $\dfrac{b-a}{b}$

    Answer: b).

16. $\dfrac{x-a}{x^2+a^2} = 0$

    a) a^3 b) $\dfrac{1-a}{a^2}$ c) $a(x^2+a^2)$ d) a e) no solution

    Answer: d).

17. $ax + x = b$

    a) $\dfrac{b}{a}$ b) $\dfrac{a+1}{b}$ c) $\dfrac{b-x}{a}$ d) $\dfrac{b}{a+1}$ e) $b(a+1)$

    Answer: d).

18. $\dfrac{x-a}{x} = b$

a) $\dfrac{a}{1-b}$ b) $\dfrac{1-b}{a}$ c) $a+bx$ d) a e) $a+b$

Answer: a).

19. $\dfrac{a}{b+\dfrac{c}{x}} = d$

a) $\dfrac{cd}{a-bd}$ b) $\dfrac{c}{ad-b}$ c) $\dfrac{(b+c)d}{a}$ d) $d-a-b-c$ e) $\dfrac{a}{d(b+c)}$

Answer: a).

20. $\dfrac{a-bx}{c-x} = d$

a) $\dfrac{a}{cd(b-1)}$ b) $a-cd-b-1$ c) $\dfrac{(b+c)d}{a}$ d) $\dfrac{a-cd}{b-d}$ e) $\dfrac{ab}{cd}$

Answer: d).

Part 3
Word Problems

Word problems are considered by many students the most difficult part of algebra. There some guidelines to follow.

1. Read the problem carefully several times.

2. Ask yourself, What is to be found? What is the question?

3. From your experience, find a formula or basic fact you can use to solve the problem.

4. Using the equation or fact, write an equation and solve it.

5. Check your answer.

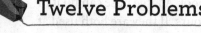

Twelve Problems

1. One number is four less than a second number. Their sum is 24. What are the numbers? Answer: 10, 14

2. Five times a number is 12 more than three times the same number. What is the number? Answer: 6

3. A man has 20 coins consisting of dimes and quarters. The total value of the coins is $3.20. How many of each coin does he have? Answer: 8 quarters and 12 dimes

4. A man has $100,000. He invests part of it at 6% interest per year. The remainder he loans to his brother at 4% interest per year. After one year, his brother has paid off the loan. His total gain on both investments is $5,200. How much did he loan to his brother? Answer: $40,000

5. A company has 500 gallons of wine containing only 10% alcohol. They decide to add some wine containing 16% alcohol to raise the level to 12%. How much of the 16% wine should be added? Answer: 250 gallons

6. Interstate 5 is a very straight road. From a given point, a car travels north, and a truck travels south. The car's speed is 50% greater than the truck's. After two hours, they are 250 miles apart. What is the speed of each vehicle? How far did each vehicle travel? Answer: The truck's rate is 50 mph, and it has gone 100 miles; the car's rate is 75 mph, and it has gone 150 miles.

7. An automobile radiator contains 6 quarts of water and 2 quarts of antifreeze. The owner wants to remove some of the mixture and replace it with pure antifreeze, to bring the radiator to at least 1/2 antifreeze. How much should he remove and replace with pure antifreeze? Answer: At least 2 2/3 quarts.

8. Two hoses are used to fill a swimming pool. A large hose can fill it in 4 hours. A smaller hose can do it in 6 hours. If they are both used at the same time, how long would it take them to fill the pool? Answer: 2.4 hours.

9. The length of a rectangle is 3 feet more than its width. If its perimeter is 42 feet, what is its length? Answer: 12 feet.

10. A woman drives from city A to city B. For the first half of the trip, due to traffic, she averages only 40 miles per hour. If she wants to average 50 miles per hour for the whole trip, how fast must she average for the second half? Answer: 66 2/3 miles per hour.

11. A rectangular swimming pool is eight feet longer than it is wide. If its area is 425 square feet, what is the length and width? Answer: 25 feet long and 17 feet wide.

12. Here is a problem similar to problem 10, but it has an unusual answer. A man drives from city A to city B. For the first half of the trip, he averages only 20 miles per hour. If he wants to average 40 miles per hour for the whole trip, how fast must he average for the second half? (You might guess 60 mph but that is wrong.) Answer: No solution.

Solutions of Twelve Problems

1. One number is four less than a second number. Their sum is 24. What are the numbers?

   Let x be the second number; then $x-4$ is the first number. The sum is $x + (x-4) = 24$. $2x-4 = 24$, $2x = 28$, $x = 14$, which is the second number. Then the first is $x-4 = 10$.

2. Five times a number is 12 more than three times the same number. What is the number?

   $5x = 3x+12$, $5x-3x = 12$, $2x = 12$, $x = 6$.

3. A man has 20 coins consisting of dimes and quarters. The total value of the coins is $3.20. How many of each coin does he have?

   Let x be the number of quarters. Then $20-x$ is the number of dimes. Since each quarter is 25 cents and each dime is 10 cents, the equation is $25x + 10(20-x) = 320$ cents.

   $25x + 200 - 10x = 320$, $25x - 10x = 320 - 200$, $15x = 120$, $x = 120/15 = 8$, $20-x = 12$.

   The number of quarters is 8; the number of dimes is 12.

4. A man has $100,000. He invests part of it at 6% interest per year. The remainder he loans his brother at 4% interest per year. After one year, his brother has paid off the loan. His total gain on both investments is $5,200. How much did he loan to his brother?

   Let x be the amount invested at 6%. Then
   $.06x + .04(100,000-x) = 5,200$,
   $.06x + 4,000 - .04x = 5,200$
   $.06x - .04x = 5,200 - 4,000$,

$.02x = 1,200,$

   $x = 1,200/.02 = \$60,000 =$ the amount invested at 6%.

Therefore, he loaned $100,000 - 60,000 = \$40,000$ to his brother.

5. A company has 500 gallons of wine containing only 10% alcohol. They decide to add some wine containing 16% alcohol to raise the level to 12%. How much of the 16% wine should be added?

Write an equation for the amount of alcohol in the wine, letting x be the amount of 16% added.

$.10(500) + .16x = .12(500+x),$

   $50 + .16x = 60 + .12x,$

   $.16x - .12x = 60-50,$

      $.04x = 10,$

         $x = 10/.04 = 250$ gallons of 16% wine.

6. Interstate 5 is a very straight road. From a given point, a car travels north and a truck travels south. The car's speed is 50% greater than the truck's. After two hours, they are 250 miles apart. What is the speed of each vehicle? How far did each vehicle travel?

Let x be the truck's speed. Then $1.5x$ is the car's speed. Then, using distance equals rate times time, $d = rt$, after 2 hours, the total distance traveled is 250 miles. $2x + 2(1.5x) = 250$ miles. $2x+3x = 250$, $5x = 250$, $x = 250/5 = 50$ mph. This is the truck's speed. Then $1.5x = 1.5(50) = 75$ mph is the car's speed. In two hours, the truck traveled $2(50) = 100$ miles. The car traveled $2(75) = 150$ miles. As a check, the total miles is $100 + 150 = 250$.

7. An automobile radiator contains 6 quarts of water and 2 quarts of antifreeze, for a total of 8 quarts. The owner wants to remove some of the mixture and replace it with pure antifreeze, to bring the radiator to at least 1/2 antifreeze. How much should he remove and replace with pure antifreeze?

We write an equation for the antifreeze. Since the total capacity is 8 quarts, the ratio of antifreeze to the total is 2/8 or 1/4. So, the radiator contains 1/4 or 25% antifreeze. Therefore, $2 - x/4 + x = (1/2)8 = 4$, $2 + (3/4)x = 4$, $3x/4 = 4-2 = 2$, $3x = 2(4) = 8$, $x = 8/3 = 2$ $2/3$ quarts.

8. Two hoses are used to fill a swimming pool. A large hose can fill it in 4 hours. A smaller hose can do it in 6 hours. If they are both used at the same time, how long would it take them?

The rate of the large hose is 1/4 pool per hour. The rate of the small hose is 1/6 pool per hour. We use the formula amount = rate times time. For both hoses together, the total rate is $1/4 + 1/6 = 3/12 + 2/12 = 5/12$ pool per hour. Since the rate times the time is one pool, let t = the time to fill the pool. Then, $(5/12)t = 5t/12 = 1$, $5t = 12$, $t = 12/5 = 2$ $2/5$ hours or 2.4 hours or 2 hours 24 minutes.

9. The length of a rectangle is 3 feet more than its width. If its perimeter is 42 feet, what is its length?

The perimeter of a rectangle is 2 times the length plus 2 times the width or $P = 2L + 2W$. Therefore, letting x be the width, $2x + 2(x+3) = 42$, $2x + 2x + 6 = 42$, $4x = 42-6 = 36$, $x = 36/4 = 9$ feet. The width is 9 feet. The length is $x+3 = 9 + 3 = 12$ feet.

10. A woman drives from city A to city B. For the first half of the trip, due to traffic, she averages only 40 miles per hour. If she wants to average 50 miles per hour for the whole trip, how fast must she average for the second half?

Since distance = rate times time, $d = rt$, time is distance divided by rate, $t = d/r$.

Since we don't know the distance for the first half, we call it d. Then the time for the first half plus the time for the second half is the total time. Let s be the speed for the second half. Then $d/50$ is the time for first half, d/s is the time for the second half, and since the total distance is $2d$, $2d/50$ is the total time. Therefore, $d/40 + d/s = 2d/50 = d/25$.

We can divide both sides by d so we get $1/40 + 1/s = 1/25$, $1/s = 1/25 - 1/40 = (40-25)/[(40)(25)] = 15/1000$. That is $1/s = 15/1000$. Multiplying both sides by $1000s$ we get, $1000 = 15s$, or $15s = 1000$ so that $s = 1000/15 = 200/3 = 66\ 2/3$ miles per hour.

11. A rectangular swimming pool is eight feet longer than it is wide. If its area is 425 square feet, what is the length and width?

   Since the area is the length times the width, $A = LW$, using x for the length and x-8 for the width, we can write the equation $x(x-8) = 425$, which can be written $x^2 - 8x - 425 = 0$. This is a quadratic equation, $ax^2 + bx + c = 0$, where $a = 1$, $b = -8$, and $c = -425$.

   Therefore we can use the quadratic formula, $x = \dfrac{-b + \sqrt{b^2 - 4ac}}{2a}$.

   Substituting the numbers, we get $x = \dfrac{-(-8) + \sqrt{64 - 4 \cdot 1 \cdot (-425)}}{2 \cdot 1}$

   or $x = \dfrac{8 + \sqrt{64 + 1700}}{2 \cdot 1} = \dfrac{8 + \sqrt{1764}}{2} = \dfrac{8 + 42}{2} = \dfrac{50}{2} = 25$.

   So the length is 25 feet. Then the width is $x - 8 = 17$ feet. Note that in the quadratic formula we use the plus sign, not the minus sign, since if we use the minus sign, the answer will be negative, and we want a positive answer.

12. A man drives from city A to city B. For the first half of the trip, he averages only 20 miles per hour. If he wants to average 40 miles per hour for the whole trip, how fast must he average for the second half?

   Since we don't know the distance for the first half, we call it d. Then the time for the first half plus the time for the second half is the total time. Let s be the speed for the second half. Then $d/20$ is the time for first half, d/s is the time for the second half, and since the total distance is 2d, $2d/40$ is the total time. Therefore, $d/20 + d/s = 2d/40 = d/20$. We can divide both sides by d, so we get $1/20 + 1/s = 1/20$, $1/s = 1/20 - 1/20 = 0$ or $1/s = 0$. But $1/s$ has to be some positive number, and it can never be zero. Therefore, there is no solution. Some people would say that the speed must be infinite.

Part 4

Subscripted Variables and the Summation Convention

Sometimes we want to refer to variables as the first x, the second x, the third x, etc. We do this with the notation

$$x_1, x_2, x_3, \ldots$$

For example, we can refer to the nth even number by $x_n = 2n$ and the nth odd number by $y_n = 2n-1$. The third even number would then be $x_3 = 2 \cdot 3 = 6$, and the fourth odd number would be $y_4 = 2 \cdot 4 - 1 = 7$.

We can see that this is true by listing the even numbers, 2, 4, 6, 8, ... The third one is obviously 6. If we list the odd numbers, 1, 3, 5, 7, 9, ... the fourth one is obviously 7.

We can also write equations such as the nth even number is one more that the nth odd number, $x_n = y_n + 1$, which is obviously true, since $2n = (2n-1)+1$.

Series and the Summation Convention

The use of a series of numbers is common in mathematics. Subscripted variables are common in the use of series. For example, $x_1 + x_2 + x_3$ would represent a series of three x's. There is also a special notation for writing series. It uses the summation notation, which is a capital Greek letter sigma, Σ which can be read as *sum of*. For example, $\sum_{i=1}^{i=3} x_i$ means $x_1 + x_2 + x_3$, *the sum of x_i from 1 to 3*. Also, it is understood that $\sum_1^3 x_i$ means the same thing, the sum on i being implied.

As an example, if we want to write the sum of the squares of the first four odd numbers, we can write it as, $\sum_1^4 (2n-1)^2$, where the sum is on n. This represents $1^2 + 3^2 + 5^2 + 7^2$.

Part 5
The Geometric Series

The annuity formula, and others derived from it in a similar way, are probably the most useful formulas studied in college, in the sense that it will be used the most in banking.

The formula is derived from the geometric series. A geometric series is one in which each term is the previous term times a constant. We call the constant r, also called the common ratio. If a_n is the nth term, then a_{n+1} is the next term. This can be said mathematically as

$$a_{n+1} = ra_n \tag{1}.$$

If a is the first term, the series is

$$S_n = \sum_{k=1}^{n} ar^{k-1} = a + ar + ar^2 + ar^3 + \cdots + ar^{n-1} \tag{2}.$$

Here, n is the number of terms in the series. We can regard formulas (1) or (2) as the definition of the geometric series.

There is a shortcut formula for the series (2) above. It is $S_n = a\dfrac{r^n - 1}{r - 1}$ (3).

Proof:

We prove the shortcut formula as follows. Equation (2) is

$$S_n = \sum_{k=1}^{n} ar^{k-1} = a + ar + ar^2 + ar^3 + \cdots + ar^{n-1} \tag{2}.$$

Multiply both sides of this by r, giving

$$rS_n = \sum_{k=1}^{n} ar^{k} = ar + ar^2 + ar^3 + \cdots + ar^{n} \tag{4}.$$

Now subtract equation (2) from (4):

$$rS_n - S_n = ar + ar^2 + ar^3 + \cdots + ar^{n-1} + ar^n - (a + ar + ar^2 + ar^3 + \cdots + ar^{n-1})$$

$$rS_n - S_n = ar + ar^2 + ar^3 + \cdots + ar^{n-1} + ar^n - a - ar - ar^2 + ar^3 - \cdots - ar^{n-1})$$

and we notice that all terms except two cancel out, leaving

$$rS_n - S_n = ar^n - a$$

or factoring we have

$$S_n(r-1) = a(r^n - 1).$$

Now dividing both sides by r-1 gives

$$S_n = a\frac{r^n - 1}{r - 1} \qquad r \neq 1,$$

which completes the proof. r cannot be 1, since then we would have a zero denominator. If $r = 1$, go back to the original series, we see that $r^n = 1^n = 1$ so all powers of r are 1. Then $S_n = a + a + a + \cdots + a = na$. This series is not very useful.

If $|r| < 1$, a more convenient form, which is the same formula, is

$$S_n = a\frac{1 - r^n}{1 - r},$$

which can be obtained by multiplying the numerator and denominator of equation (3) by -1.

As a simple example, let $a = 1$, $r = 2$, and $n = 4$, so

$$S_4 = 1 + 2 + 2^2 + 2^3 = 1 + 2 + 4 + 8 = 15;$$

and to check the formula,

$$S_4 = 1\frac{2^4 - 1}{2 - 1} = \frac{16 - 1}{1} = 15.$$

Problem: Calculate $S_4 = 1 + 3 + 3^2 + 3^3$ by adding and check it from the shortcut formula. Answer: 40.

Part 6
Saving Money and the Annuity Formula

In the problems below, you will need a scientific calculator to do the more complex calculations. The y^x key on most scientific calculators is useful to compute any number to any power.

Now we derive two formulas, one for saving money and one for paying off a loan.

The formulas are derived from the geometric series we discussed above. The formulas for the geometric series we repeat here:

$$S_n = \sum_1^n ar^{k-1} = a + ar + ar^2 + \cdots + ar^{n-1} \tag{2},$$

where n is the number of terms. The shortcut formula for this, as derived earlier, is

$$S_n = a\frac{r^n - 1}{r - 1} \tag{3}.$$

1. Saving Money

Suppose someone saves P dollars at the end of each month for n months at the interest rate of i per month. That is, if the annual rate is 6%, then the monthly rate is $6/12 = .5\%$ or $i = .005$. As a concrete example, suppose someone saves $P = \$100$ per month for three years (36 months) at an annual interest rate of 6% per year or .5% per month.

The amount that someone has in any month is the amount he or she had in the previous month plus i times that amount, plus what he or she saved that month, P. That is,

$$S_{k+1} = S_k + iS_k + P.$$

This can be shortened slightly to

$$S_{k+1} = S_k(1+i) + P \tag{5}.$$

This is called a recursion formula, since it tells how S recurs from month to month. If we start with P at the end of the first month, then

$$S_1 = P$$

$$S_2 = S_1 + iS_1 + P = S_1(1+i) + P = P(1+i) + P = P[(1+i)+1]$$

$$S_2 = P[(1+i)+1]$$

$$S_3 = S_2(1+i) + P.$$

Now, in this formula, for S_3 substitute $S_2 = P[(1+i)+1]$:

$$S_3 = S_2(1+i) + P = P[(1+i)+1](1+i) + P = P[(1+i)^2 + (1+i) + 1].$$

Now we repeat the process with $S_4 = S_3(1+i)+P$ to get

$$S_4 = P[(1+i)^3 + (1+i)^2 + (1+i)+1].$$

Continuing in this way we finally get

$$S_n = P[(1+i)^{n-1} + (1+i)^{n-2} \cdots +(1+i)+1].$$

By reversing the order inside the brackets, we can rewrite it

$$S_n = P[1+(1+i)+(1+i)^2 +(1+i)^3 \cdots +(1+i)^{n-1}] \tag{6}.$$

Or in summation we get

$$S_n = P\sum_{k=1}^{n}(1+i)^{k-1} \tag{7}.$$

The summation is actually a geometric series with $a=1$ and $r=i+1$, so using the shortcut formula,

$$S_n = P\frac{r^n-1}{r-1} = P\frac{(1+i)^n-1}{1+i-1} = P\frac{(1+i)^n-1}{i} \tag{8}.$$

This is called the annuity formula for saving money at a constant rate. It is how much money you have after n payments. It is a very important formula for all banking transactions. Sometimes we need P given S_n, so solving this formula for P gives

$$P = S_n \frac{i}{(1+i)^n -1} \tag{6}.$$

This is the payment needed to save to save S_n in n months at a rate of i per month.

Example 1:

If you save \$100 per month for 3 years at an annual rate of 6%, how much money will you have?

Since the annual rate is 6% the monthly rate is 6/12 or .5% so $i = .005$ and $n = 36$ months.

$$S_{36} = 100\frac{(1+.005)^{36}-1}{.005} = 100\frac{1.19668-1}{.005} = 100\frac{.19668}{.005} = 100(39.3361) = 3933.61$$

(We used the y^x key to calculate $(1.005)^{36} = 1.19668$.) Therefore, you will have saved $3,933.61. Note if there were no interest, the amount would be $36(100) = \$3,600$.

Example 2:

If the annual interest rate is 6% per year, how much should you save each month if you want to have $10,000 at the end of three years?

Using $P = S_n \dfrac{i}{(1+i)^n - 1}$ we have

$$P = 10,000\frac{.005}{(1+.005)^{36}-1} = 10,000\frac{.005}{1.19668-1} = \frac{50}{.19668} = \$254.22.$$

You must save $254.22 each month.

2. Paying Off a Loan

Now suppose someone takes out a loan of A dollars and makes a payment of P dollars at the end of each month for n months at an interest rate of i per month. How much will be owed at the end of k months?

Let A_k be the amount owed at the end of the kth month. Let A be the initial loan amount. The monthly interest is added to the loan each month, and the payment is subtracted. Then the recursion formula from one month to the next is

$$A_{k+1} = A_k + iA_k - P = A_k(1+i) - P \qquad (7).$$

Therefore, after one month, the amount owed is

$$A_1 = A(1+i) - P \qquad (8).$$

And after two months

$$A_2 = A_1(1+i) - P \qquad (9).$$

If we substitute A_1 from equation (8) into equation (9), we get

$$A_2 = \left[A(1+i) - P\right](1+i) - P$$

$$A_2 = A(1+i)^2 - P\left[(1+i) + 1\right].$$

Then, substituting in a similar manner, we get

$$A_3 = A_2(1+i) - P$$

$$A_3 = \{A(1+i)^2 - P[(1+i)+1]\}(1+i) - P$$

$$A_3 = A(1+i)^3 - P\left[(1+i)^2 + (1+i) + 1\right].$$

And after k substitutions, we get

$$A_k = A(1+i)^k - P\left[(1+i)^{k-1} + \cdots + (1+i) + 1)\right].$$

By reversing the quantity in brackets, this can be written

$$A_k = A(1+i)^k - P\left[1 + (1+i) + \cdots (1+i)^{k-1}\right].$$

But the quantity in brackets is a geometric series with $a = 1$, $r = 1 + i$, so that $r - 1 = 1 + i - 1 = i$.

Using formula (3) above for the geometric series for the term in brackets, we can write it as

$$A_k = A(1+i)^k - P\frac{(1+i)^k - 1}{i} \tag{10}.$$

This is the unpaid balance after k payments.

This formula can be used to calculate the payment needed to pay off a loan. To pay off a loan in n payments, we want the unpaid balance, A_n, to be zero. Therefore,

$$A(1+i)^n - P\frac{(1+i)^n - 1}{i} = 0$$

or

$$A(1+i)^n = P\frac{(1+i)^n - 1}{i}.$$

Solving for P, we get

$$P = A(1+i)^n \frac{i}{(1+i)^n - 1} = \frac{i(1+i)^n}{(1+i)^n - 1}A.$$

By dividing top and bottom by $(1+i)^n$, this simplifies to

$$P = \frac{i}{1-(1+i)^{-n}}A \tag{11}.$$

You can calculate the $(1+i)^{-n}$ term by using the y^x key on your scientific calculator.

An interesting calculation is to find how much is paid for interest only. Since you paid n payments of P dollars each, the total amount paid is nP. Therefore, the amount paid in interest only is that amount minus the amount of the loan you paid off. That is, if B is the amount paid in interest only,

$$B = nP - A \qquad\qquad\qquad (12).$$

Example 3:

A person buys a house and takes out a loan of $500,000 for 30 years (360 months) at 6% annual interest. What will be her house payment, and how much did she pay in interest only? $A = 500,000$, $i = .005$ and $n = 360$, so that the payment is

$$P = \frac{i}{1-(1+i)^{-n}} A = \frac{.005}{1-(1+.005)^{-360}} 500,000 = \frac{.005}{1-.16604} 500,000 = \frac{.005}{.83396} 500,000$$

$P = .00599(500000) = \$2997.75$. This is the monthly payment.

The amount paid in interest only is

$B = nP - A = 360(2997.75)\text{-}500,000 = 1,079,190 \text{ - } 500,000$
$B = \$579,190.$

You may be shocked to know that you paid more in interest than the amount of the loan!

Summary of Annuity Formulas

1. If P is saved for n months at a rate of i per month, we have the amount saved is

$$S_n = P\frac{(1+i)^n - 1}{i}.$$

2. If you want to have to have amount S_n dollars saved after n months, and if the interest rate is i per month, the amount you want to save per month is

$$P = \frac{i}{(1+i)^n - 1} S_n.$$

3. In paying off a loan of amount A, the unpaid balance after k months will be

$$A_k = A(1+i)^k - P\frac{(1+i)^k - 1}{i}.$$

The monthly payment to pay off the loan in n months is

$$P = \frac{i}{1-(1+i)^{-n}} A.$$

4. The total amount paid in interest only is $B = nP - A$.

Annuity Formula Problems

1. If you save $250 per month for 5 years at an annual interest rate of 9%, how much will you then have?   ANSWER: $18,856.03

2. If you want to have $50,000 in 4 years at an annual rate of 9%, how much should you save per month?   ANSWER: $869.25

3. If you borrow $600,000 to buy a house at an annual rate of 9% for 15 years,

    a). What is your monthly payment?   ANSWER: $6,085.60

    b). How much did you pay in interest?   ANSWER: $495,408.00

Solutions to Annuity Formula Problems

1. If you save $250 per month at an annual interest rate of 9%, how much will you then have after 5 years? We use the formula below where $i = .09/12 = .0075$ and $n = 12 \cdot 5 = 60$ months.

$$S_n = P\frac{(1+i)^n - 1}{i} \qquad S_{60} = 250\frac{(1+.0075)^{60} - 1}{.0075}$$

$$S_{60} = 250\frac{1.56568 - 1}{.0075} = 250 \cdot (75.42414) = \$18,856.03$$

2. If you want to have $50,000 in 4 years at an annual rate of 9%, how much should you save per month? $i = .09/12 = .0075$, $n = 4 \cdot 12 = 48$ months. Here, $S_{60} = S_{48} = 50000$. We use the formula

$$P = \frac{i}{(1+i)^n - 1} S_{60}$$

$$P = \frac{.0075}{(1.0075)^{48} - 1} 50,000 = \frac{.0075}{1.43141 - 1} 50,000 = \$869.25.$$

3. If you borrow $600,000 to buy a house at an annual rate of 9% for 15 years,

   a). What is your monthly payment? $A = 600,000$, $n = 15 \cdot 12 = 180$ months, and $i = .09/12 = .0075$. We use the formula

   $$P = \frac{i}{1-(1+i)^{-n}} A$$

   $$P = \frac{.0075}{1-(1.0075)^{-180}} 600,000 = \frac{.0075}{1-.26055} 600,000 = \$6085.60.$$

   b). How much did you pay in interest?

   $$B = nP - A = 180(6085.60) - 600,000 = \$1,095,408 - 600,000 = \$495,408$$

Appendix A
Why You Can't Divide by Zero

Argument 1

We give an indirect proof. That is, to prove statement P is true, we assume the opposite, P is false, or not-P is true. If the opposite, not-P, leads to a contradiction, then not-P is false, which the means its opposite, the original statement P, is true.

To prove P, "you can't divide by zero," we suppose not-P, "you can divide by zero." That is, suppose 1/0 were some real number. Call that number x. That is,

$$x = \frac{1}{0}.$$

But we know from algebra that if $x = \frac{1}{b}$, then we can multiply both sides by b to get $bx = 1$. But if $b = 0$, then we get $0 \cdot x = 1$ or $0 = 1$ which is obviously false. This is a contradiction, so the original assumption that "$x = \frac{1}{0}$ is a number" is false. Therefore, P, "you can't divide by zero," is true.

Argument 2

Let us see what happens when we let x, in the expression $1/x$, get closer and closer to zero.

$$\frac{1}{.1} = 10, \ \frac{1}{.01} = 100, \ \frac{1}{.001} = 1,000, \ \frac{1}{.0001} = 10,000.$$

We see that as x approaches zero, then $1/x$ gets larger and larger. So that we can say that as x approaches zero, $1/x$ approaches infinity so x can never be zero.

Appendix B
Why Minus Times Minus Is Equal to Plus

We give two separate arguments, one involving torque and one using algebra.

Torque Argument

We will answer this question by considering an application from the subject of mechanics, a branch of physics. There is a quantity called *torque*. If you have ever used a wrench or screwdriver or opened a jar of pickles, you have used torque, T. It is defined as force, F, times distance, D. The unit of torque is foot-pounds or inch-pounds. The formula is as follows:

$$T = FD$$

Let us consider a wrench loosening or tightening a nut. We will define the force F as positive up and negative down. We define the distance D as positive if to the right of the nut and negative if to the left. We define torque as positive counterclockwise (CCW) and negative clockwise (CW). Consider the wrench shown below:

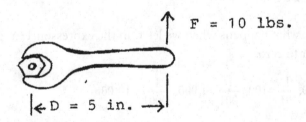

For this figure, $T = FD = 10 \cdot 5 = 50$ inch-pounds. Here, F and D are positive and the nut turns counterclockwise (CCW), so the torque is positive.

Now consider the following two figures:

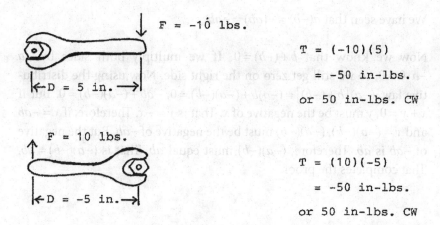

In the first figure, the force is down, negative, but the distance is still to the right, positive. One can see that the nut rotates clockwise, so the torque is negative. In the second figure, the force is up, positive, but the distance is to the left of the nut, so it is negative. The the nut still rotates clockwise, so the torque is still negative.

Finally, consider the figure below.

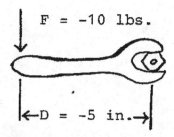

Here the force is down, negative, and the distance is to the left, also negative. But the nut rotates counterclockwise, which we have agreed makes the torque positive.

T = (-10)(-5) = +50 inch-pounds.

This means that it is natural to define *minus times minus equal to plus.*

Algebra Argument

We have seen that $a(-b) = -(ab) = -ab$.

Now we know that $b + (-b) = 0$. If we multiply both sides by $-a$ $-a \cdot 0 = 0$, so we still get zero on the right side. Now using the distributive law, $(-a)[b + (-b)] = (-a)b + (-a)(-b) = 0$, $-ab + (-a)(-b) = 0$. But if $x + y = 0$, y must be the negative of x. That is $y = -x$. Therefore, if $x = -ab$ and $y = (-a)(-b)$, $(-a)(-b)$ must be the negative of $-ab$. But the negative of $-ab$ is ab. Therefore, $(-a)(-b)$ must equal ab. That is $(-a)(-b) = ab$. That completes the proof.